SpringerWienNewYork

CISM COURSES AND LECTURES

THIN FILMS OF SOFT MATTER

EDITED BY

SERAFIM KALLIADASIS
IMPERIAL COLLEGE LONDON, UK

UWE THIELE
MAX-PLANCK- INSTITUT FÜR PHYSIK KOMPLEXER SYSTEME,
DRESDEN, GERMANY

SpringerWien NewYork

PREFACE

This book contains lecture notes, albeit not covering all material, delivered at the Advanced Course 'Thin Films of Soft Matter' that took place at CISM Udine in July 18-22, 2005.

Thin film flows of soft matter (either simple Newtonian liquids or polymeric and other complex materials) are often encountered in a wide variety of natural phenomena and technological applications: from gravity currents under water and lava flows to heat and mass transport processes in conventional engineering applications and more recent developments in the area of nanotechnology and MEMS. In the vast majority of cases, thin film flows are bounded by either free surfaces which separate the film from the surrounding phases, as in the case of jets or soap films, or by a free surface and a solid substrate. The involved scales range from the nanometer level as for dewetting thin polymer films and break-up of nanojets to the centimetre scale as for heat and mass transport applications to the meter scale as for lava flows.

The Course aimed at giving a detailed overview of the main and most up-to-date advances in the area of thin films and jets, through a balanced combination of theory and experiments. Since the subject is essentially an interdisciplinary area and as such it brings together scientists and engineers with different educational backgrounds, it was important to offer a research-oriented exposition of the fundamentals of free-surface flows in confined geometries. The goal was to arrive at ready-to-use mathematical models of different degrees of complexity which are capable of describing accurately thin film and jet flows in a relatively 'simple' (i.e. avoiding using the full Navier-Stokes equations) and experimentally testable way.

A wide range of topics was covered: basic equations and interfacial boundary conditions, as well as derivation of model equations for the evolution of the free surface including long-wave expansions and equations of the boundary-layer type; linear stability analyses, weakly and strongly nonlinear analyses including construction of stationary periodic, solitary wave and similarity solutions; interfacial instabilities and formation of complex wave structures; dewetting on chemically homogeneous and heterogeneous substrates; influence of surface tension gradients due to the thermal Marangoni effect and thermocapillary Marangoni instabilities; rupture/dewetting for very thin one- and two-layer films; miscible fingering in electrokinetic flow as a model system to study extended domain dynamics problems, such as the falling film problem, driven by the dominant zero modes associated with key symmetries; influence of chemical reactions and three-dimensional effects on falling liquid films; generic treatment of self-similarity, scaling laws, dimensional analysis and scaling theory of singularities, which is crucial not only for the specificity of the Course but also from the point of view of general mechanics; singularity formation and topological transitions such as drop break-up and nanojet break-up; experimental characterization of capillarity such as spreading drops, wetting of textured surfaces, wicking and coating; experimental characterization of thin films using atomic force microscopy, ellipsometry and contact angle measurements, and analysis of patterns using Minkowski functionals.

The Course was organized at the suggestion of Professor Manuel G. Velarde, while he was in office as Rector of CISM (2002-2005). It was not a mere suggestion. He provided the organizers (now the editors of this book) with valuable recommendations about topics, names of potential lecturers and an important advice: emphasis on combination of theory, natural phenomena, lab experiments and numerical experiments. This was the natural thing for he had been and still is engaged in the field covered by the Summer School. On the other hand both the present editors had been and still are collaborating with him in some of the themes discussed here. Though Rector of CISM he participated as a student in our School.

Professor Velarde was born in Almeria, Spain, September 7, 1941 and hence he is turning 65 in 2006. He has made seminal and long lasting contributions to fluid physics and, in particular, to our understanding of hydrodynamic instabilities and convective pattern dynamics, wetting and spreading dynamics, interfacial hydrodynamics, and nonlinear wave dynamics. In the latter subfield he is responsible for introducing the concept of 'dissipative soliton' to account for experiments on surface tension gradient-driven waves. In spite of the diversity and heterogeneity of his publications, Professor Velarde has always emphasized the unifying perspective underlying them all. This he taught us to look for in our own research.

We the editors take pleasure in dedicating this book to Professor Manuel G. Velarde on his 65th birthday, with the hope of seeing him again as a student in another of our courses while continuing collaborating with him, a superb teacher.

Manuel G. Velarde amicitiae et admirationis ergo.

The coordinators of the School and editors of this book wish to express their appreciation to Professors Hsueh-Chia Chang, Jens Eggers, David Quéré and Ralf Seemann for their acceptance of both duties, lecturing and writing lecture notes. They specially thank Ioan Vancea for help with the final typesetting. They also acknowledge the CISM Scientific Council for their encouragement and support. Last but not least, they are grateful to the Secretariat staff of CISM and in particular Signora Elsa Venir and Signora Carla Toros for their efficient handling of administrative matters before and during the duration of the School.

Serafim Kalliadasis (London), Uwe Thiele (Dresden)

CONTENTS

Structure Formation in Thin Liquid Films: Interface Forces Unleashed

Ralf Seemann,[*] Stephan Herminghaus,[*] Karin Jacobs[‡]

[*] Max Planck Institute for Dynamics and Self-Organization, Göttingen, Germany
[‡] Experimental Physics, Saarland University, Saarbrücken, Germany

Abstract We present a conclusive overview of the stability conditions and the dewetting scenarios of thin liquid coatings. The stability of thin films is given by the effective interface potential $\phi(h)$ of the system and depends among other parameters on the film thickness h. In the case of unstable or metastable films holes will appear in the formerly uniform layer and the film dewets the substrate. We describe the analysis of emerging hole patterns and how to distinguish between different dewetting scenarios. From this analysis we derive the effective interface potential for our particular system, $\phi(h)$, which agrees quantitatively with what is computed from the optical properties of the system. Our studies on thin polystyrene films on Si wafers of variable Si oxide layer thickness demonstrate that the assumption of additivity of dispersion potentials in multilayer systems yields good results and are also in accordance with recent numerical simulations.

Thin liquid films on solid substrates are present in everyday life, e.g. as lubricant film on the cornea of our eyes or on the piston in a car's motor, but also as ink on a transparency. In some cases, these films are not stable on their substrate and bead up, a phenomenon that is easily observable if one tries to paint an oily surface. In recent years, much effort has been put into understanding the dewetting phenomena in thin films on solid substrates, both experimentally (Reiter, 1992; Bischof et al., 1996; Xie et al., 1998; Jacobs et al., 1998a; Herminghaus et al., 1998; Kim et al., 1999; Sferrazza et al., 1998; Seemann et al., 2001b; Tsui et al., 2003; Seemann et al., 2005; Fetzer et al., 2005) and theoretically (Vrij, 1966; Ruckenstein and Jain, 1974; Brochard-Wyart and Daillant, 1989; Sharma and Khanna, 1998, 1999; Konnur et al., 2000; Koplik and Banavar, 2000; Thiele et al., 2001, 2002; Thiele, 2003; Sharma, 2003; Münch, 2005). On the one hand, industry is interested in the prevention of dewetting in order to achieve e.g. stable lithographic resists. On the other hand, basic research is still demanding general rules to infer the rupture and dewetting mechanisms from the spatial ordering and timely evolution of the dewetting pattern. The knowledge of the underlying forces and mechanisms would enable us to predict stability conditions for practical use.

Earlier studies of dewetting scenarios either concentrated on the morphological characterization (Reiter, 1992; Bischof et al., 1996; Xie et al., 1998; Jacobs et al., 1998a) or on the simulation of dewetting patterns on the basis of an assumed effective interface potential (Konnur et al., 2000; Koplik and Banavar, 2000). Some studies took long-range forces into account, but faced the difficulty that their origin and strength was not known precisely (Herminghaus et al., 1998; Kim et al., 1999; Sferrazza et al., 1998). We have therefore studied a model system polystyrene (PS) on silicon (Si) substrates with various silicon oxide (SiO) layers which accounts for all of the important features observed in more complex situations. Owing to the simplicity of the model system we used, a thin film (< 60 nm) of a non-polar liquid on a solid substrate, it is possible to calculate quantitatively the effective interface potential (Israelachvili, 1992) and to compare it to what is derived from the dewetting morphology.

To study dewetting phenomena, a system was used that is on the one hand close to application (coatings, photoresist), yet on the other hand easily controllable in the experiments. Polymers such as polystyrene (PS) are very suitable model liquids since they have a very low vapor pressure in the melt, and mass conservation is valid. Moreover, they are chemically inert, non-polar, and their dynamics can be tailored by choosing different chain lengths and annealing temperatures. Below the glass transition temperature T_g, the film is glassy and can be stored for subsequent analysis. For this work, atactic polystyrene was used due to the known absence of any crystallization in this material.

If not denoted otherwise thin polystyrene films ('PS(2k)', $M_w = 2.05$ kg/mol, $M_w/M_n = 1.05$, Polymer Labs, Church Stratton, UK) were prepared from toluene solution onto three types of silicon wafers: Type A and B with thin natural oxide layer of 1.7 nm or 2.4 nm respectively (Wacker Chemitronics, Burghausen, Germany; (100)-oriented, p-doped, conductivity < 10 Ωcm) and type C with a thick oxide layer of 190(1) nm (Silchem GmbH, Freiberg, Germany; (100)-oriented, p-doped, conductivity > 1 Ωcm). Prior to coating, the silicon wafers were degreased by sonicating them in ethanol, acetone, and toluene. Residual hydrocarbons were etched away by a 30 min dip in fresh 1:1 H_2SO_4 (conc.)/H_2O_2 (30%) solution. Subsequently, the acids were removed by a thorough rinse in hot Millipore^{TM} water.

Cleaning and coating were performed in a class 100 clean room. The thicknesses of the silicon wafers' oxide layer and of the polymer films were measured by ellipsometry (Optrel GdBR, Berlin, Germany). In what follows, we use the term 'SiO' for the silicon oxide layer, despite the fact that most of the amorphous layer consists of silicon dioxide (Sze, 1981). Further characterization of wafers and polymer films was done by atomic force microscopy (AFM) (Multimode III, Digital Instruments, Santa Barbara) using Tapping Mode^{TM}. AFM revealed the rms-roughness of the silicon wafers to be below 0.2 nm, that of the polymer film below 0.3 nm. Samples were annealed on a temperature-controlled hot plate to temperatures between 50 and 140°C for typically 2 to 360 min. AFM scanning parameters were optimized not to affect the liquid polymer films. In some cases, also X-ray diffraction (grazing-incidence diffraction at Troika II, ESRF, Grenoble) was used to determine PS film thickness.

To start with, it is necessary to clarify the distinction between stable, metastable, and unstable films. This is straightforward in terms of the effective interface potential, $\phi(h)$, which is defined as the excess free energy (per unit area) it takes to bring two interfaces from infinity to a certain distance, h (Dietrich, 1988; Schick, 1989). In our case, the two interfaces involved are the solid/liquid interface and the liquid/air interface, and h_0 is the initial thickness of the liquid film. By definition, $\phi \to 0$ for $h \to \infty$, as shown for three important cases in Figure 1. The solid line (1) characterizes a film that is stable on the substrate, since energy would be necessary to thin the film. The equilibrium film thickness is infinite. The two other curves exhibit a global minimum of $\phi(h)$ at $h = h_{eq}$ and the system can gain energy by changing its present film thickness h to h_{eq}.

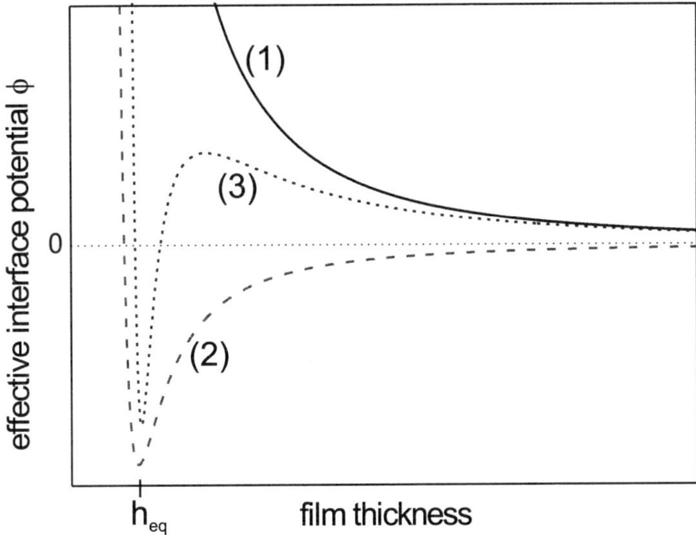

Figure 1. Sketch of the effective interface potential $\phi(h)$ as function of film thickness for a stable (1), unstable (2), and metastable film (3), respectively.

Dewetting is characterized by the formation of circular holes, their growth in time and coalescence, finally leading to a set of droplets on the substrate. A negative global minimum of $\phi(h)$ at a finite film thickness h_{eq} indicates that after dewetting of the film into droplets, an equilibrium layer of thickness h^* will be left on the substrate. The depth of that minimum is connected with the contact angle θ_{eq} of the liquid on the solid substrate (Dietrich, 1988; Schick, 1989; Frumkin, 1938; Seemann et al., 2001d):

$$|\phi(h_{eq})| = \sigma(1 - \cos\theta); \tag{1}$$

where σ is the surface tension of the liquid air interface.

A typical dewetting experiment under lab conditions is shown in Figure 2. Here, two main rupture mechanisms are possible. i) Dry spots are nucleated. Nucleation may be initiated by defects as, e.g., dust particles (heterogeneous nucleation) or by thermal nucleation. ii) Capillary waves are spontaneously amplified. The latter mechanism is called 'spinodal dewetting' (Vrij, 1966; Ruckenstein and Jain, 1974; Brochard-Wyart and Daillant, 1989). It is readily shown that spinodal dewetting can take place only if the second derivative of ϕ with respect to film thickness is negative, $\phi''(h_0) < 0$, where h_0 is the initial thickness of the homogeneous film. Whenever this is the case, the system is called unstable. Hence, the dashed curve (2) of Figure 1 characterizes an unstable film. The dotted curve (3) describes a film that is unstable for small film thicknesses, where $\phi''(h) < 0$, whereas for larger film thicknesses, only nucleation can drive the system towards dewetting. Here, the film is called metastable.

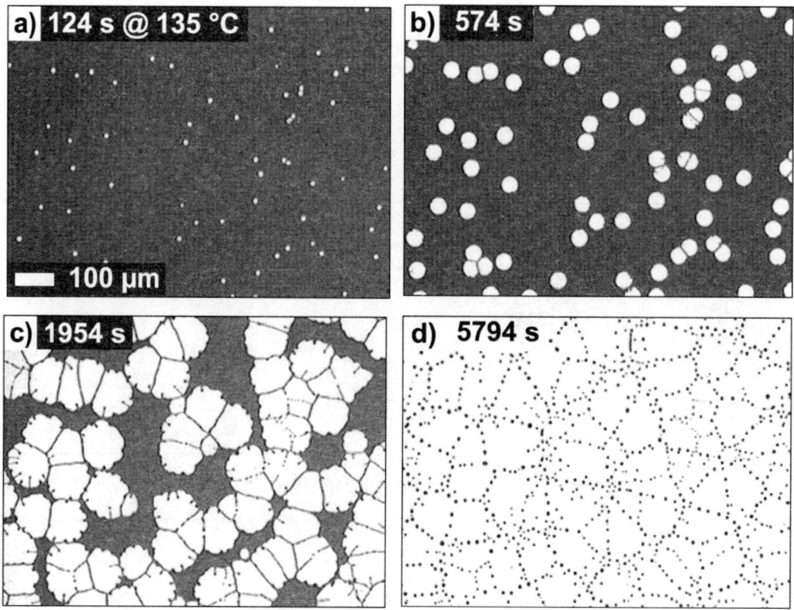

Figure 2. Pictures taken by a light microscope: a 80 nm thick polystyrene film of 65 kg/mol molecular weight is dewetting at 135°C from a hydrophobized silicon substrate. a) t = 2 min, b) t = 10 min, c) t = 33 min, d) t = 97 min.

The clue to the effective interface potential is its connection to the characteristic features of the dewetting pattern. In case $\phi''(h)$ is negative, all fluctuations in film thickness with wavelengths above a certain critical threshold λ_c are amplified and grow exponentially according to $\exp(t/\tau)$, where τ is the growth time that is characteristic for the respective mode which leads to spontaneous dewetting of the liquid (spinodal dewetting). There is, however, a certain wavelength λ_s, the amplitude of which grows fastest, leading to a characteristic dewetting pattern of the liquid film. This spinodal wavelength λ_s is experimentally observable and is linked to the effective interface potential (Vrij, 1966; Ruckenstein and Jain, 1974; Williams and Davis, 1982; Brochard-Wyart and Daillant, 1989):

$$\lambda_s(h) = \sqrt{\frac{-8\pi^2\sigma}{\phi''(h)}}. \tag{2}$$

This process is analogous to spinodal decomposition of a blend of incompatible liquids which occurs if the second derivative of the free energy with respect to the composition is negative. There, as well, a certain wavelength exists the amplitude of which grows the fastest. Following this analogy, dewetting via unstable surface waves has been termed 'spinodal dewetting' (Mitlin, 1993). Equation (2) illustrates that only if $\phi''(h) < 0$ (spinodal dewetting film), λ_s is real. For $\phi''(h) = 0$, $\lambda_s(h)$ diverges to infinity.

In the metastable case, for a film thickness where $\phi''(h) > 0$, the system has to overcome a potential barrier in order to reach its state of lowest energy at $h = h_{eq}$. Some kind of nucleus, e.g. a dust particle, is required to lower $\phi(h)$ and can therefore induce dewetting. This rupture mechanism is termed 'heterogeneous nucleation' (Mitlin, 1993, 1994). Close to the sign reversal of $\phi''(h)$ it is called 'homogeneous nucleation'. In this case, no nucleus is necessary because the thermal 'activation' is sufficient to overcome the energy barrier (Blossey, 1995).

Equation (2) indicates, that spinodally dewetting films contain experimentally accessible information about the underlying forces and are therefore of special interest. But to make use of equation (2), we first have to prepare a spinodally dewetting film. Our first aim therefore is to unambiguously recognize a spinodally dewetting film. Theoretically, the distinction between nucleation and spinodal dewetting was quite clear: Vrij (1966) proposed already in 1966 that a spinodal rupture of a free liquid film results in a dewetting pattern of 'hills and gullies' with a preferred distance λ_s after a certain time of rupture τ. Experimentally, the rupture time τ is difficult to measure since the hole must have a certain size to be observable. Experimentalists instead concentrated on the evidence of a preferred wavelength λ_s in their systems.

If, however, the holes are randomly (Poisson) distributed, they are assumed to stem from heterogeneous nucleation, reflecting the fact that nuclei typically exhibit random statistics. But in a real experimental system we face a couple of difficulties: the thicker the films are, the weaker is the driving force, and the larger is the growth time τ of the spinodal mode (typically, $\tau \propto h^5$). For thicker films, dewetting by heterogeneous nucleation may therefore be quicker and can suppress a spinodal pattern (Konnur et al., 2000; Becker et al., 2003; Thiele et al., 2001, 2002). Moreover, chemical heterogeneities locally cause a change in ϕ and therefore the rupture conditions of the sample may vary from spot to spot leading to a less ordered dewetting pattern. This effect is more pronounced in thicker films due to the small driving forces and the large growth time τ.

Hence, due to the fact that both nucleation and spinodal dewetting might lead to dewetting in unstable films, the experimental distinction between spinodal dewetting and heterogeneous nucleation is far from being obvious: Experiments referring to the scaling of the preferred hole distance via the spinodal wavelength ($\lambda_s \propto h^2$) and therefore just inferring the average hole density as function of film thickness will very probably fail.

For the distinction between the mechanisms we therefore probe the statistics of the distribution of the sites of the holes. We first determine the simple two-point correlation function $g(r)$ of the point set represented by the positions of the centers of the holes arising in a dewetting polystyrene film on a hydrophobized silicon substrate, Figure 3(a). Obviously, no feature that indicates a dominant wavelength can be found.

Let us consider for comparison the dewetting pattern of a liquid gold film on a quartz glass, as shown in Figure 3(b) and determine $g(r)$, too. Again, no modulation in $g(r)$ can be detected. For this system, however, spinodal dewetting with a dominant wavelength λ_s had been clearly identified as the dewetting mechanism (Bischof, 1996).

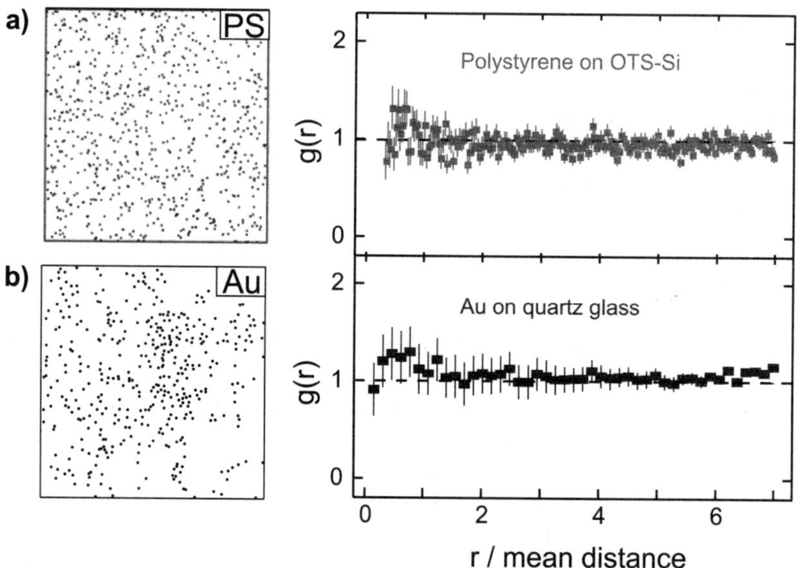

Figure 3. a) Positions of holes in a PS film (left), as extracted from an optical micrograph and the corresponding pair correlation function $g(r)$. b) Position of hole sites in a Au film as extracted from an atomic force micrograph, and as well the corresponding g(r). r is given in units of the mean distance of the holes.

For dewetting patterns where the preferred wavelength cannot be detected by a radial pair correlation function g(r) or a Fourier transform (before the film rupture), more powerful tools have to be applied. Here, Minkowski functionals - based on integral geometrical methods - have shown to be a versatile method to track down higher order correlations (Mecke, 1994; Herminghaus et al., 1998; Jacobs et al., 1998a). Their application to the experimental system of this work is described in detail by Jacobs et al. (2000) in a comparative study on dewetting patterns of gold and of PS films.

The central idea is to determine the spatial statistics of the hole positions by adding a morphology to the point pattern (Mecke, 1994). This is done by assigning circular disks of radius r to each of the hole positions, as shown in Figure 4(a). Due to possible overlap of the disks, the area F and the boundary length U of the set union of disks do not increase proportional with r^2 or r, respectively. The larger the overlap, the slower will be the increase of the two measures. A third Minkowski functional, the Euler characteristics χ, is a measure for the connectivity of a pattern and is defined in two dimensions as the mean curvature of the boundary line. For a random set of points, the dependence on r of the three Minkowski functionals was analytically determined and then compared with the results obtained for the holes in the gold and in the PS films. The behavior of the Minkowski measures for the holes in the PS films was shown to be in accordance with the result for a random set of holes (Jacobs et al., 1998a); see Figure 4(b). The Minkowski measures for the holes in the gold film, however, differed greatly from the curve of a 'random' sample, demonstrating the presence of correlations between the sites of the holes. This meant that the precondition for a spinodal mechanism was matched for the gold films, but not for PS films on the hydrophobized Si substrates in that study.

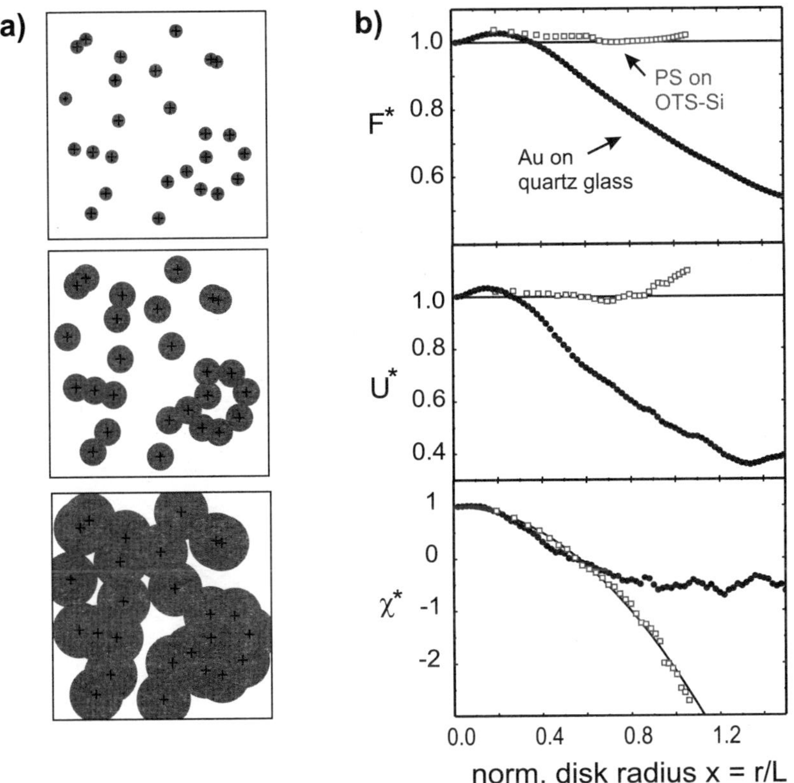

Figure 4. a) The positions of the holes, marked with a cross, are decorated each with a disk, whose radius increases from left to right. The Minkowski functionals in two dimensions include area F (the grey area), boundary length U between grey and white area and the Euler characteristic χ, which is a measure of the connectivity of the grey structure. b) Normalized morphological measures F^*, U^* and χ^* of the Au (full circles) and of the PS film (open squares) as a function of the normalized radius x, $x = 3D\, r/L$, of the disks with mean distance L. The solid lines mark the expected behavior for a Poisson point process.

The method of determining the Minkowski measures of a point set was applied in all our studies, whenever the presence of correlations between the sites of the holes could not be shown by a Fourier transform or a radial pair correlation function.

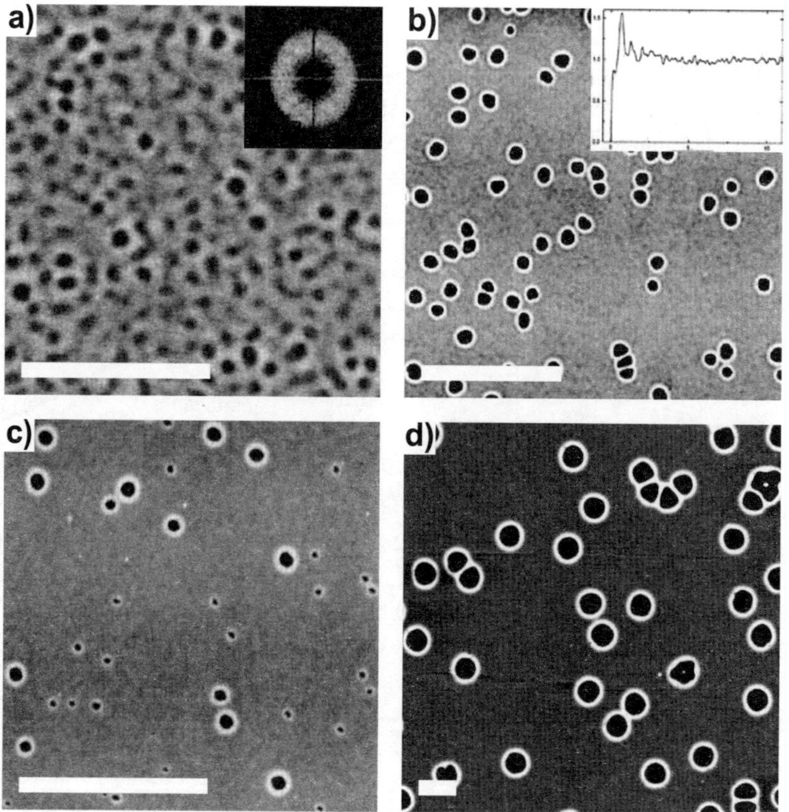

Figure 5. Dewetting patterns of PS(2k) films as seen by AFM; the height scale is ranging from black (0 nm) to white (10 nm; for d) 20 nm). The length bar denotes 5 μm. a) PS film on type C wafer with thickness of 3.9(2) nm, b) 3.9(2) nm PS on type B wafer, c) 4.1(1) nm PS on type B wafer d) 6.6(4) nm PS on type B wafer. Patterns shown in a) and b) are typical for spinodal dewetting, the pattern in c) is characteristic for homogeneous (thermal) nucleation and the pattern in d) stands for a typical scenario of heterogeneous nucleation. (The statistical analysis of the distribution of hole sites in cases b) to d) was performed on larger sample areas.)

In Figure 5, typical snapshots of dewetting patterns are shown. We classify the dewetting patterns into four categories, a representative of each is shown in Figure 2. For the classification we take into account correlations between the sites of the troughs of the undulation: a pattern shown in Figure 5(a) is described as 'densely packed crests and troughs with preferred distance', in Figure 5(b) as 'correlated holes within a uniform film matrix', in Figure 5(c) as 'randomly distributed holes of different sizes', and in Figure 5(d) as 'randomly distributed holes of one size'.

In Figure 5(a) a 3.9(2) nm thick PS(2k) film dewets from a type C wafer. Many crests and troughs cover the entire film. Some of the troughs are in fact holes since they already touch the substrate. No more uniform film is detectable. A Fourier transform reveals a preferred distance of heaps or bumps, respectively, of 340(30) nm (Seemann et al., 2001a). X-ray diffraction measurements at a later stage of dewetting indicate that the holes are not 'dry'. Rather, they are covered with a residual PS film of thickness 1.3(2) nm. Such a dewetting pattern can be detected up to some 10 nm. For thicker films on that substrate, the time τ until the pattern is sufficiently clear to be observed steeply rises from seconds to months with increasing film thickness ($\tau \propto h^5$). During very long annealing times, however, holes nucleated by e.g. dust particles grow rapidly in size and the entire film is 'eaten up' by those holes before the above described spinodal pattern can develop (Sharma and Khanna, 1998, 1999).

Figures 5(b-d) show a PS(2k) film dewetting from a type B wafer, with increasing film thickness from left to right. Up to a film thickness of 3.9(2) nm, we again observe a preferred distance of holes, but there is still uniform film surrounding the dewetted spots, as shown in Figure 5(b). As compared to PS films on type C wafers, we measure on type B wafers larger preferred distances of holes for the same film thickness. Here, we discover correlations in hole sites either by a ring in the Fourier transform of the image or by observing a modulation within the radial pair correlation function calculated for the sites of holes (Xie et al., 1998; Jacobs et al., 2000; Seemann et al., 2001a). With only slightly thicker PS films, $h_o = 4.1(1)$ nm, the most striking feature is that we observe holes of different sizes within one AFM image and that more and more holes pop up in the course of the experiment, cf. Figure 5(c). Moreover, within the experimental error bar, no correlations of hole sites can be detected. For films of 6 nm thickness and larger, we find a dewetting scenario exemplarily shown in Figure 5(d). Here, we observe isolated circular holes of about identical radius (variance is less than 5%). Upon longer annealing times, these holes grow, but no additional holes emerge.

A detailed analysis with the help of Minkowski functionals reveals that the holes are randomly (Poisson) distributed, as described in an earlier study (Bischof et al., 1996). Up to now, the very nature of the nucleation sites has not been satisfactorily revealed. Already in 1979 Croll found that due to the preparation of thin polymer films from solvents the macromolecular chains are not in an equilibrium state. He found that the stress in a polystyrene film cast from toluene solution at about 20 °C is, quite universally, 14 MPa (Croll, 1979). By annealing the films on a wettable substrate (and thereby most likely reducing the stress inside the films) prior to the transfer to the non-wettable samples, we were able to show that a certain fraction of holes can be suppressed (Podzimek et al., 2001). This has recently been corroborated by Reiter et al. (2005).

By increasing the prepared polymer film thickness on waver type B and C, the dominant wavelength λ_s increases, too, as expected for a spinodal dewetting scenario (Vrij, 1966). Experimental data of $\lambda_s(h)$ are shown in Figure 6(a) as the filled squares. A preferred wavelength λ_s can also be found for PS(2k) films on type B wafers, but only for film thicknesses smaller than 4.1 nm (open circles in Figure 6(a)). On type A and B wafers, PS(2k) films smaller than 3.2 nm could not be prepared by spin coating from toluene solution, they dewetted during the spin coating process.

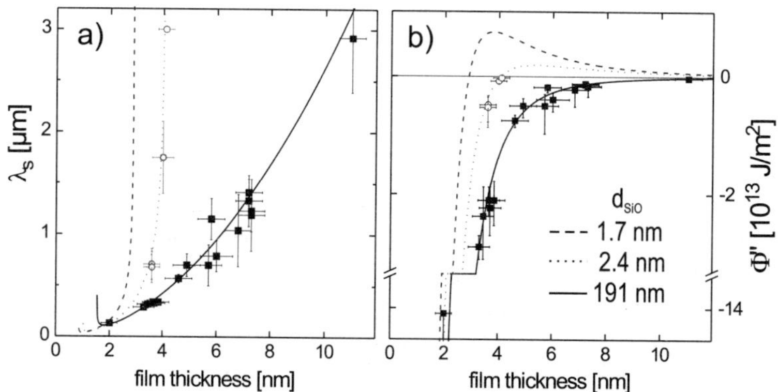

Figure 6. a) Spinodal wavelength λ_s as function of PS film thickness h on type B (open circles) and type C wafers (filled squares). b) Second derivative of effective interface potential ϕ'' as function of film thickness h. Note the axis break of ϕ''.

With the help of equation (2), and using $\sigma = 30.8$ mN/m, we can infer $\phi''(h)$ from $\lambda_s(h)$, as shown in Figure 6(b). In case of a PS film on a type C wafer, a simple van der Waals potential of the form

$$\phi_{vdW}(h) = -A_{SiO}/(12\pi h^2), \tag{3}$$

can be fitted to the data of $\phi''(h)$ (Figure 6(b), solid line). A_{SiO} is the Hamaker constant for a PS film on a SiO substrate. Here, A_{SiO} is a fit parameter with the value $A_{SiO} = 2.2(4)$ E^{-20} J, which is in excellent agreement with the Hamaker constant calculated by refractive indices and dielectric constants of the materials in the layered system air/PS/SiO (Israelachvili, 1992): $A_{SiO,calc} = 1.8(3)$ E^{-20} J.

Since a van der Waals potential, as shown in equation (3), cannot explain a global minimum of $\phi(h)$ at the equilibrium film thickness h_{eq}, we introduce a model potential that additionally contains a $1/h^8$ dependence, which is one of the models commonly used in this context:

$$\phi(h) = c/h^8 + \phi_{vdW}(h). \tag{4}$$

Here, the first term includes short range interactions of strength c and the second term characterizes the long-range interactions by the van der Waals potential.

$\phi(h)$ has to fulfill two further conditions which define the position and depth of the global minimum, $\phi(h_{eq})$. First, the contact angle θ of the liquid on the substrate, which can be determined independently by AFM (7.5° in our case), fixes the depth of the global minimum, equation (1). Second, the position h_{eq} of the global minimum is determined by the equilibrium film thickness of 1.3(2)nm as measured by X-ray reflectivity. We found that by a suitable choice of the free parameter c, both conditions can be fulfilled. We obtain $c = 6.3(1)$ E^{-76} Jm^6 for a 191 nm SiO layer. The reconstructed effective interface potential is plotted as solid line in Figure 3. For all experimental film thicknesses (< 60 nm) $\phi''(h)$ is negative and therefore the PS films on type C wafers are unstable.

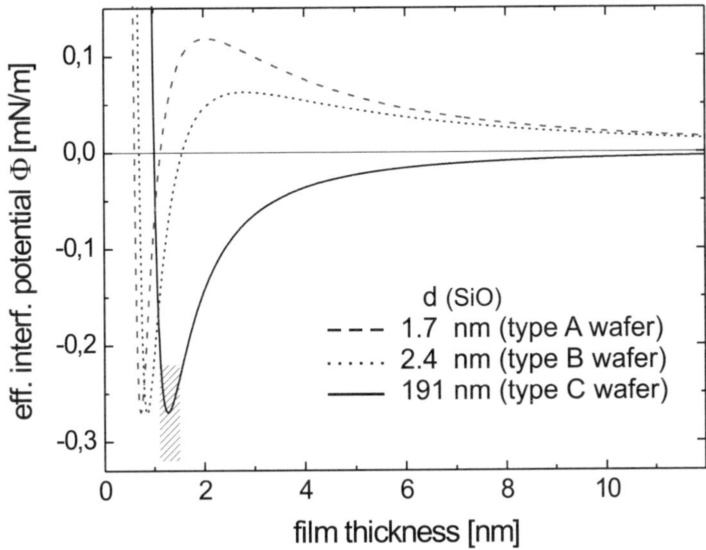

Figure 7. Reconstructed effective interface potential $\phi(h)$ for polystyrene films on three types of silicon wafers. The hatched rectangle indicates in x-direction the experimental error for h_{eq} from the X-ray measurements, and in y-direction the error in $\phi(h_{eq})$ due to the limited accuracy of the contact angle measurement.

Up to now, we treated the PS films on type C wafers as a 'two interfaces'-system, air/PS/SiO, and neglected the bulk silicon underneath the SiO. For type A and B wafers, however, with oxide layer thicknesses of only 1.7 nm and 2.4 nm, respectively, this approximation turns out not to be adequate. These have to be treated as 'three interfaces' systems, air/PS/SiO/Si. This may be done by considering the van der Waals potential of the PS film to be composed of two terms: one of the SiO layer and one of the silicon bulk corrected by the SiO layer thickness d:

$$\phi_{vdW}(h) = -\frac{A_{SiO}}{12\pi h^2} + \frac{A_{SiO} - A_{Si}}{12\pi(h+d)^2}. \tag{5}$$

It is one of the aims of the discussion below to test the adequacy of this procedure.

In case of type B wafers, we now apply the same procedure for reconstructing $\phi(h)$ as before, but use equation (5) instead of equation (3) for $\phi_{vdW}(h)$. By fitting the second derivative of equation (5) to the experimental data for $\phi''(h)$ in Figure 6(b) and using the Hamaker constant A_{SiO} from above, we obtain $A_{Si} = $ -1.3(6) E^{-19} J. That again corresponds very well to the calculated value of $A_{Si,calc}= $ -2.2(5) E^{-19} J. Following the procedure as described before, we obtain a strength c of the short range attraction of c = 5.1(1) E^{-77} Jm^6 for a 2.4 nm SiO layer. $\phi(h)$ is plotted as dotted curve in Figure 7. For films thicker than 4.1 nm, $\phi''(h)$ is positive and thus dewetting will only proceed by nucleation, whereas for thinner films spinodal dewetting is possible. Both are consistent with our experimental observation on type B wafers.

For PS films on type A wafers, we can now reconstruct $\lambda_s(h)$ even without any experimental data points for λ_s: the long-range part of $\phi(h)$ can be calculated by equation (5) with $d = 1.7$ nm and the Hamaker constants mentioned before. The short-range part can be obtained by relating the experimental contact angle θ to the global minimum of ϕ via equation (4) and choosing the strength c adequately. We find $c = 1.8(1)$ E^{-77} Jm^6. Here again, $\phi(h)$ describes a metastable PS film (cf. Figure 7, dashed line) and $\phi''(h)$ is positive for PS films thicker than 2.9 nm, which is consistent with our experiments, where we observe only nucleated holes for films thicker than 3.2 nm. This is different with earlier work using silicon wafers with a native oxide layer as a substrate (Xie et al., 1998). In that work, patterns reminiscent of spinodal dewetting were observed for PS thicknesses up to 10 nm. According to our findings, this can never be achieved with a native oxide layer. Furthermore, no agreement was reached between the calculated and experimental Hamaker constant.

From the reconstructed effective interface potentials for all three types of substrates we can infer the particular dependence of the spinodal wavelength λ_s on PS film thickness h via equation (2). All three $\lambda_s(h)$ curves are plotted in Figure 6(a). The results of these experiments are summarized in Figure 8. Films that exhibit patterns with a structure factor are marked as open symbols. Systems with only randomly distributed holes are given solid symbols. A star marks a situation as described in Figure 5(c), where more and more holes pop up in the course of the experiment. (The single star-shaped point in Figure 8 comprises three data points lying on top of each other.) The three types of wafers are identified by triangles (type A), circles (type B), and squares (type C). These experimental results can be understood in terms of the reconstructed effective interface potential $\phi(h)$ (Seemann et al., 2001a), that is shown in Figure 7.

From the potentials we can infer the thickness $\tilde{h}$ below which $\phi''(h)$ is negative and spinodal dewetting is possible: for type A wafers, $\tilde{h} = 2.9(3)$ nm and for type B wafers, $\tilde{h} = 4.1(3)$ nm. Here, the error bars account for the errors determining the oxide layer thicknesses and the Hamaker constants involved. For type C wafers, $\tilde{h} = 325$ nm as determined from the potential shown in equation (5). Accounting again for the error in $\phi(h)$ due to the involved Hamaker constants, $\tilde{h}$ is within the interval 230 nm $< \tilde{h} <$ 495 nm. Experimentally, this relatively large error interval is nevertheless irrelevant, since an estimation of the spinodal rupture time τ of an even 230 nm thick PS(2k) film on type C wafer gives values for τ larger than 2000 years [1]. In other words, spinodal dewetting on Si/SiO substrates is observable under lab conditions for PS film thicknesses below 15 nm only. Otherwise, as described before, heterogeneous nucleation will always be faster in initiating holes.

---

[1] For the estimation of τ we used equation (31) of Sferrazza et al. (1998), $\tau = 48\pi h\sigma h^5/A^2$ with viscosity $h = 3200$ Pa s, surface tension $\sigma = 31$ mN/m and a Hamaker constant A of SiO$_x$ of $A = 2.2$ E-20 J, taken from Seemann et al. (2001a).

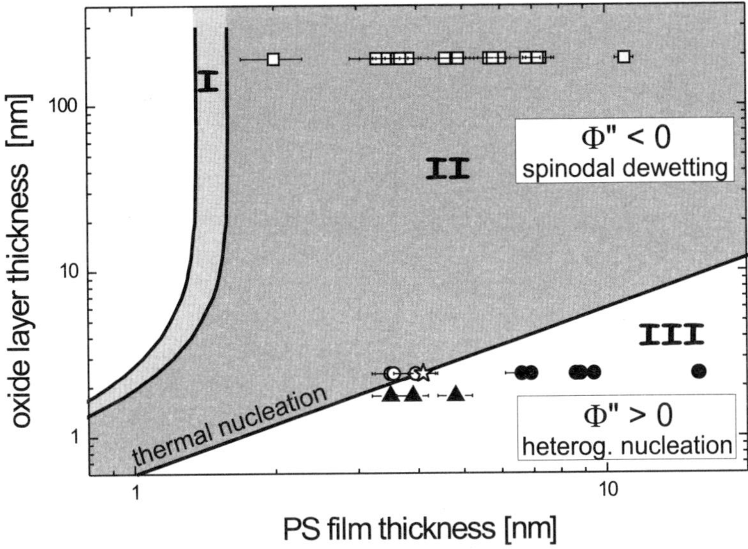

Figure 8. Stability diagram of PS films on top of silicon wafers with variable oxide layer thickness. Experiments that exhibit patterns with a structure factor (e.g. Figure 5(a) and (b) are marked as open symbols. Systems with only randomly (Poisson) distributed holes (e.g. Figure 5(d)) are given solid symbols. A star marks a situation as described in Figure 5(c). The three types of wafers used in the experiments are identified by triangles (type A), circles (type B), and squares (type C). In the grey areas, regimes I and II, PS films can dewet by a spinodal mechanism, in regime III only heterogenenous nucleation is possible.

With the help of the equation (5) for the effective interface potential $\phi(h)$, we are now able to predict stability conditions of PS films on Si/SiO substrates: From the reconstructed $\phi(h)$ we can infer the zero of $\phi''(h)$ as a function of oxide layer and PS film thickness. This curve is plotted in Figure 8 [2]. It hence separates the spinodal regime II (dark gray area) from the regime III of heterogeneous nucleation (bright area) and may be called 'spinodal line' [3]. Along the spinodal line, homogeneous nucleation by thermal fluctuations is possible (Blossey, 1995). The region of thermal nucleation is very narrow, effectively collapsing to a single line in this graph. For silicon wafers without any oxide layer, equation (5) predicts PS films to be stable. This, too, is found experimentally

---

[2]It should be pointed out that in the unstable regimes I and II as well as in the metastable regime III, heterogeneous nucleation is possible and indeed is observed as explained later in the text, but spinodal dewetting can only take place in regimes I and II.

[3]The spinodal line here is analogous to the spinodal line separating the metastable and the spinodal (unstable) region in the phase diagram of the decomposition of two incompatible liquids.

(Stange et al., 1997; Jacobs et al., 1998b), if the oxide layer is stripped away (e.g. by an HF dip) prior to PS film preparation.

Let us now compare the experimental data points plotted in Figure 8 with the regimes predicted by the effective interface potential of equation (2). In regime II, data points with open symbols are found, which account for the fact that a spinodal wavelength λ_s was observed and measured. These data of λ_s were used to reconstruct the interface potentials shown in Figure 7. It is therefore obvious that in regime II only open symbols are found. Filled symbols stand for experiments were randomly dispersed holes were detected. From these experiments, we cannot reconstruct ϕ. Nevertheless, all of the filled data points, measured on type A and B wafers, are located in regime III. Hence, the reconstructed effective interface potentials correctly predict the dewetting behavior even in these cases. Of particular interest in this context are the experiments on type B wafers. The stability diagram in Figure 8 predicts that for $h < \tilde{h} = 4.1(1)$ nm spinodal dewetting is possible, whereas for thicker films, nucleation is the only dewetting mechanism. This exactly corroborates our experimental findings. Close to the sign reversal of $\phi''(h)$ at $h = \tilde{h} = 4.1(1)$ nm, thermal nucleation is predicted. Indeed, typical signs of thermal nucleation are experimentally observed: holes are generated during the entire dewetting process and the sites of the holes are not correlated. Hence, on type B wafers, all three types of rupture mechanisms are theoretically predicted and experimentally observed.

It is now challenging to compare our results with simulations of dewetting patterns of Sharma and Khanna (1998, 1999). They proposed that the morphology of the dewetting pattern depends on the form of the effective interface potential at the present film thickness h, or, more precisely, on the course of $\phi''(h)$. To explain this in more detail, the effective interface potential $\phi(h)$ and its second derivative $\phi''(h)$ are sketched in Figure 9 for an unstable (a) and for a metastable film (b) as function of film thickness h. For both systems, the global minimum of $\phi(h)$ is at finite film thickness h_{eq}. We divided the course of the effective interface potential now into three regimes, following a suggestion of Sharma and Khanna (1998, 1999). Regimes II and III are defined as before in Figure 8 and differ in the sign of $\phi''(h)$. In regime I, $\phi''(h)$ is negative, too, and the film can spinodally dewet, but $|\phi''(h)|$ is decreasing with decreasing film thickness, whereas in regime II it is increasing. The results of the simulations of Sharma and Khanna in regime I and II are shown in Figure 9(c) and 9(d), respectively. In regime I, Figure 9(c), the formerly uniform film has been transformed into a landscape of only crests and troughs, whereas in regime II, Figure 9(d), isolated holes can be found within a matrix of still uniform film. Sharma and Khanna explain this considerable difference in behavior by the course of $|\phi''(h)|$, which acts as a kind of driving force for the instability. In regime II, deep troughs (meaning sites of small PS film thickness) experience a stronger driving force than shallow ones, whereas it is vice versa in regime I. Therefore, in regime II single holes in a film matrix are prevailing, yet in regime II shallow troughs also may lead to holes, resulting in a 'crests and troughs' pattern morphology. In regime III, where only nucleation is possible, the holes in the film will be distributed according to the statistics of the nuclei. In Figure 9(e), a Poisson distribution of 40 nuclei or holes, respectively, is depicted.

To compare qualitatively our experimental dewetting patterns with the ones of Sharma and Khanna's simulations, we assign a pattern like the one in Figure 5(a) with a simu-

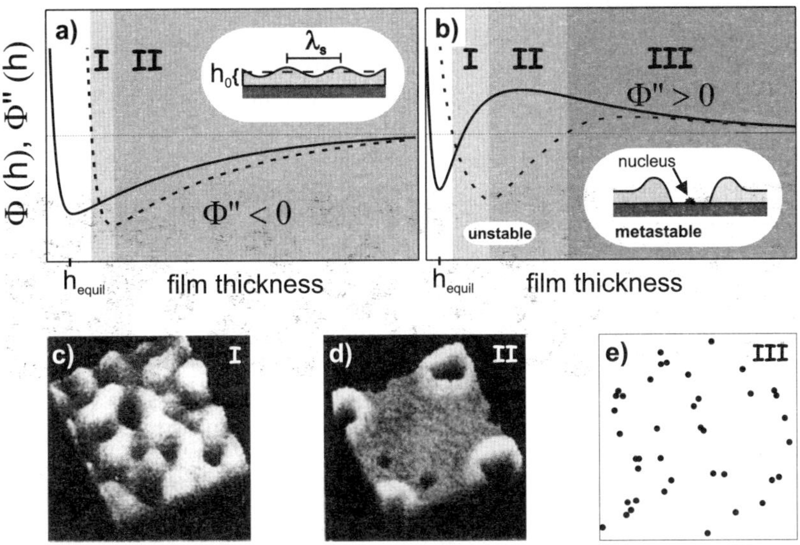

Figure 9. a) and b): Sketch of the effective interface potential $\phi(h)$ (solid line) and $\phi''(h)$ (dashed line) for a) an unstable and b) a metastable system. The insets illustrate the respective dewetting mechanism. c) and d): Simulations of dewetting patterns in regime I (c), regime II (d), pictures taken from Sharma and Khanna (1998, 1999). e) Illustration of a random (Poisson) distribution of 40 holes.

lated pattern depicted in Figure 9(c). In both patterns, no uniform film area seems to be left. A pattern consisting of isolated holes like the one shown in Figure 5(b) may hence correspond to the simulated pattern of Figure 9(d). Since we know the course of $\phi(h)$ and $\phi''(h)$ for our experimental system, we are able to check whether the experiment takes place in regime I, II, or even III. In Figure 8 we therefore also plotted the zero passage of $\phi'''(h)$ as a function of oxide layer and PS film thickness. It separates regime I (light gray area) from regime II (dark gray area). Since all of our experimental data points are located in regime II or III, respectively, and none in regime I, the above assignment of the patterns depicted in Figure 5(a) and Figure 9(c) is obviously not quite correct. But because a clear threshold behavior is not expected at this point we can at least state that the image Figure 5(a) is very close at the boundary to regime I. To understand this issue, we examine in the following the entire dewetting process, from the onset of an undulation to the equilibrium state of the PS film, a set of droplets on the substrate, and check for the predicted temporal evolution of the undulating spinodal pattern.

Figure 10 depicts the entire structural evolution of a 3.9 nm PS(2k) film on a type C wafer. The sample is monitored by AFM in tapping mode^{TM} while being annealed. AFM scans were continuously recorded, whereby one scan of $(1.5\ \mu\mathrm{m})^2$ took 1 min. Up to about 5000 s, the temperature was held constant at 53 °C. Afterwards, the temperature was successively increased up to 100 °C in order to reach the equilibrium state (only droplets on the substrate) within less than three hours. In Figure 10 we therefore note down annealing times only for scans taking place at T = 53 °C. In Figures 10 (a-d), the AFM scan area is $(1.5\ \mu m)^2$, whereas in Figures 10 (e-g) it is $(8\ \mu m)^2$. These three larger scans, made in between the series of small scans, exhibit no trace of any kind of square pattern of $(1.5\ \mu m)^2$ size and hence reassure that the liquid dewetting film has not been damaged by continuous scanning.

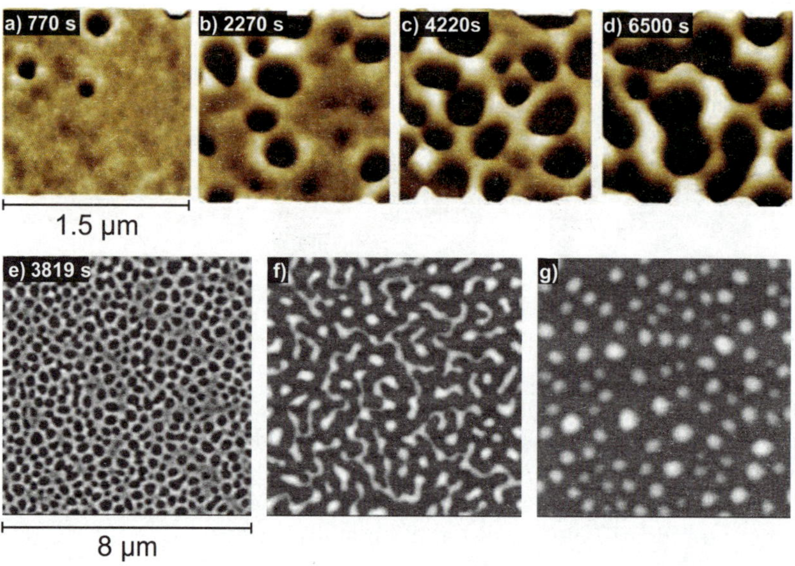

Figure 10. Dewetting morphology of a 3.9(2) nm PS(2k) film on a type C wafer as recorded by *in situ* AFM tapping mode^{TM}. Up to about 5000 s, the temperature was held constant at 53 °C and annealing times at start of the measurement are given in the pictures. Afterwards, the temperature was successively increased to 100 °C. Scan size in a)-d) is $(1.5\ \mu m)^2$. In e)-g) control scans were made of $(8\ \mu m)^2$ to check possible damage of the sample by the AFM tip. Scan g) characterizes the end of the dewetting process.

As the AFM scans reveal, dewetting proceeds in this system by the appearance of an undulation on the surface of the PS film, the troughs of which later lead to circular holes. Here, deep troughs lead to holes at earlier times than shallow troughs. Let us resume the comparison of the morphologies resulting from experiment and simulation, respectively. We stated earlier that a pattern like the one in Figure 5(a) or Figure 10(e) consists of only crests and troughs and could be assigned to the simulated dewetting pattern of regime I. According to oxide layer and PS film thickness of the system, however, the sample should be assigned to regime II. The *in situ* AFM scans yet reveal that even a 'crests and trough' pattern morphology as shown in Figure 10(e) may stem from isolated holes that pop up within a narrow time window. Comparing the dewetting pattern of an earlier AFM snapshot, e.g Figure 10(a), with the simulated patterns rather suggests an assignment to regime II. In other words, judging the regimes by a single snapshot of a dewetting simulation or of a dewetting experiment can be misleading, since also the temporal development of the pattern must be taken into consideration. Our experiments hence corroborate the results of Sharma and Khanna, who claim that the form of the potentials determines the morphology of the dewetting pattern. In experiments with PS(2k) on Si and SiO wafers, however, regime I is hardly accessible since it requires extremely small PS film thicknesses.

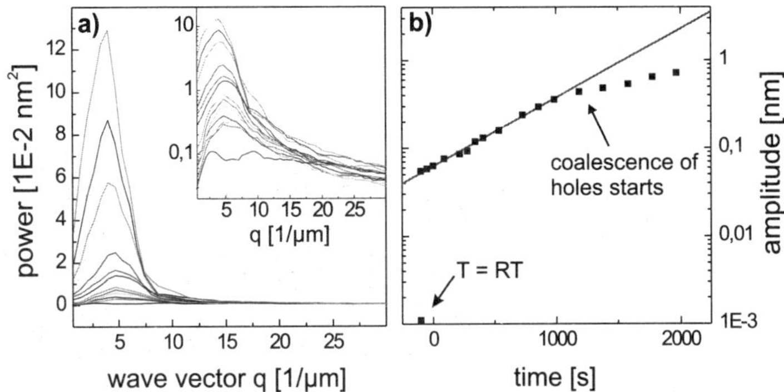

Figure 11. Results of Fourier transforms of *in situ* AFM pictures at T = 53 °C, some of which are shown in Figure 10. a) Power spectral density plot of the Fourier transforms. In the inset, a semilog plot is shown. Here, the lowest curve reveals that prior to annealing, no preferred wave vector is observed. Note that the time intervals between the curves are not constant (see right). b) Amplitude of the undulation as function of annealing time. The first data point at t < 0 s gives the roughness of the PS film surface at room temperature (RT) as revealed from a Fourier transform. The solid line is a fit of an exponential growth to the data.

Since it is possible with AFM to follow the spinodal dewetting process in situ, we can determine the spinodal wavelength as function of time via a Fourier transform. A radial average of the Fourier transform is shown in Figure 11(a) for several AFM scans of the early stages of dewetting, i.e. up to about 1000 s, when only the undulation and very small holes are present. For annealing times larger than about 1000 s, holes start to coalesce and the growth of the amplitude slows down. In Figure 11(b), the amplitude of the preferred wavelength is shown as function of annealing time. As seen by the semilog plot before hole coalescence, the amplitude grows exponentially with time and can therefore be described in the early stage of dewetting by $\exp(t/\tau_{fit})$. A fit to the data yields $\tau_{fit} = 560(20)$ s. τ_{fit} can be regarded as an upper limit for the rupture time, since it was obtained by averaging over the undulation present on an $(1.5\ \mu m)^2$ area. Nevertheless, τ_{fit} can be compared on the one hand with the time the first holes break up, which can be estimated from the AFM scans to about 230(50) s. τ scales linearly with the viscosity η as $\tau = (48\pi^2\sigma h^2\eta)/A^2$ (Ruckenstein and Jain, 1974). If surface tension σ, film thickness h and Hamaker constant A are known, the viscosity can be inferred directly. For instance, the viscosity of $\eta \approx 12{,}000$ Pa s of the a 3.9 nm thick PS(2k) film at $T = 53°$C as determined from τ is about six orders of magnitude lower than the bulk viscosity.

Such a low viscosity implies also that in these thin films T_G must be lowered. Viscosity and T_G can be related via the Vogel-Fulcher law. Data for T_G obtained via determining τ are in full agreement with data obtained by optical dilatometry (ellipsometry) for the same system and complete the latter to smaller film thicknesses. Both data sets can be described by the same fitting function (Herminghaus et al., 2001, 2003; Herminghaus, 2002). Moreover, these results are in agreement with complementary experiments measuring the molecular motion of PS(2k) molecules with nuclear magnetic resonance (Herminghaus et al., 2004; Seemann et al., 2006)

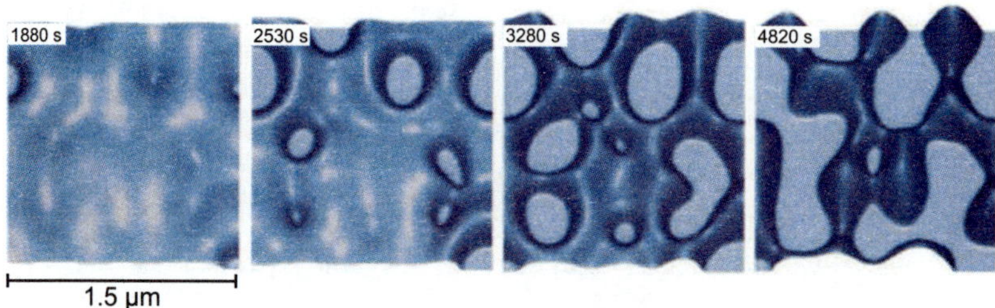

Figure 12. Dewetting morphology of a thin film. Images taken from Becker et al. (2003). Simulated spinodal dewetting morphology with the identical system parameters as in the experiment of Figure 10. The highest points reach 12 nm above the bottom of the holes. The simulations started with a slightly perturbed film.

Since it was possible to reconstruct the effective interface potential ϕ and the viscosity η of the experimental system, it is now challenging to perform numerical simulations. This was done in a collaboration with theoretical physicists and applied mathematicians (Becker et al., 2003; Neto et al., 2003a,b). For the simulations, experimental input was given by the reconstructed function ϕ, as described before and by film thickness h, surface tension σ, and viscosity η of the liquid film as known for the experiments. A comparison of the temporal evolution of the dewetting morphologies in simulation, see Figure 12, and experiment, see Figure 10, shows that the experimental process can indeed be reproduced by simulations: the time scale of the simulated film rupture as well as the morphology correspond well to the experimental data. A detailed comparison of the emerging patterns in experiments and simulations was done with the help of Minkowski functionals (Becker et al., 2003). Using these functionals was a major step in the description of film rupture, creating a powerful tool to study the impact of various parameters on film rupture, an example of which will be given in the following. Very recent results now also give insight into the important role of thermal fluctuations to dewetting (Fetzer et al., 2006). Actual improvements indicate that the match of the time scales can be improved even further including thermal fluctuation of the liquid air interface (Fetzer et al., 2006).

To complete the detailed pattern analysis of spinodal dewetting, we show in Figure 13 a peculiar dewetting scenario of thin PS films, found in simulations and also in experi-

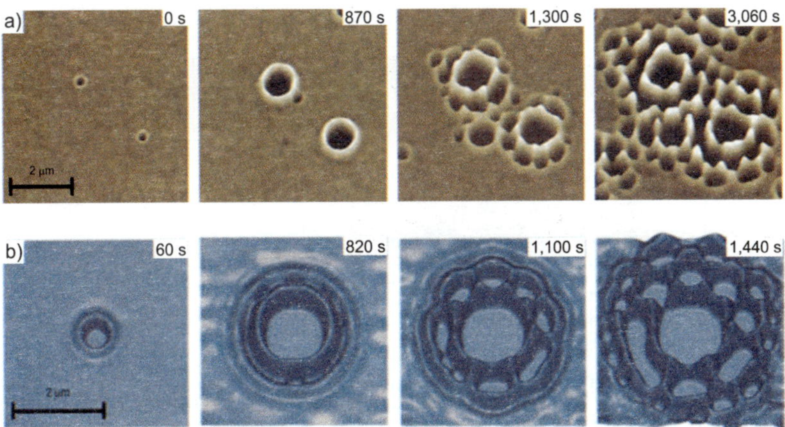

Figure 13. Dewetting morphology of a thin film. Images taken from Becker et al. (2003) a) Experiment: Temporal Series of AFM scans recorded in situ at $T = 70°C$; a 4.9 nm PS film beads off an wafer with 191 nm oxide thickness. b) Simulated dewetting morphology with the identical system parameters as in the experiment. The highest points reach 12 nm above the bottom of the holes. As initial data we took a slightly corrugated film with a depression in its center.

ments. The morphology can be understood by observing its temporal evolution. First, single holes pop up; then, along the perimeter of each hole, secondary and later ternary holes appear.

To understand this behavior we have to explore the rim profile of a dewetting film. A detailed analysis of the shape of the rim surrounding the holes revealed that under certain conditions the rim does not decrease monotonically from the crest to the prepared film thickness, but rather exhibits an oscillatory behavior (Seemann et al., 2001c), as first predicted by Thiele et al. (2001, 2002). The inset of Figure 14(a) presents an AFM scan of a typical hole in a 6.6(2) nm thick PS film on a wafer B. A trough is visible as a dark circle surrounding the wet side of the bright crest of the rim. The sketch in Figure 14(a) shows a radial cross section of this hole. The most outstanding feature of the cross section is the way the rim decays into the unperturbed film. The rim profile is asymmetric, with higher slopes near the three-phase contact line and a 'trough' on the wet side of the rim where it meets the undisturbed film. Here, the form resembles a damped harmonic oscillation. From the temporal evolution of the experimental rim profile, we can determine the height A of the rim and the depth B of the trough as a function of time. Figure 14(b) shows how $|B|$ increases for early time, respectively small rim heights A and plateaus for larger times.

Under certain conditions, the depth of the through $|B|$ does not reach a plateau value but rather grows until a new hole is nucleated (Seemann et al., 2001c; Herminghaus et al., 2002). Upon further experimental and theoretical investigation, it became clear that these 'satellite' holes in fact come from a breakthrough of the trough. A spinodally

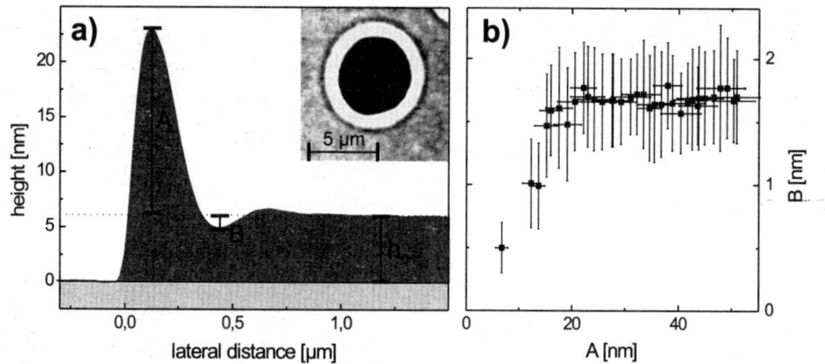

Figure 14. a) Inset: AFM scan of a hole in a 6.6(2) nm thick PS film on a wafer B after 30 min at T = 80 °C; z-scale ranges from 0 nm (black) to 20 nm (white). The large diagram shows a radial cross section of this hole, where A describes the height of the rim, B the depth of depression. b) Typical evolution of the depth of depression B over height of rim A of the hole.

unstable system strengthens the tendency for the trough to rupture (Herminghaus et al., 2002). An identical behavior was also found by Grün and Becker in their simulations (Becker, 2004; Becker et al., 2003). It should be mentioned here that some theoretical studies proposed this kind of morphology, too, but claimed that certain preconditions are necessary, e.g. that it must be an evaporative system (Kargupta et al., 2001). Our experiments and also recent simulations, however, demonstrate that this behavior should be typical for thin films of Newtonian liquids, particularly in the unstable state and within a certain thickness regime (Neto et al., 2003a,b; Herminghaus et al., 2002).

Further studies - simulations and experiments - explored the "satellite patterns" also on straight dewetting fronts and on dewetting fronts meeting under a certain angle (Neto et al., 2003a,b) There, too, experiments and simulations correspond very well.

In conclusion, we have gained a comprehensive picture of the film stability by a detailed analysis of the emerging dewetting patterns. We were able to distinguish experimentally spinodally dewetting films from homogeneous or heterogeneous nucleated ones. Studies on spinodally dewetting films allowed us to derive the effective interface potential ϕ, which is the clue for understanding the interplay of short- and long-range forces. Moreover, a simple approach of using additive van der Waals forces turned out to be adequate for correctly describing the potential. In situ AFM experiments on dewetting films additionally gave insight into the viscosity decrease of thin PS films. The reconstructed effective interface potential ϕ and the reduced viscosity η allowed for a comparison of the morphology and the dynamics of the evolving dewetting patterns in experiment and numerical simulation.

The comprehensive consistency of all characteristic dewetting features now allows the tailoring of systems according to the desired wetting properties. A future challenge is to move on to more complex systems with rough surfaces or to metal or liquid substrates. Also highly polar materials or viscoelastic fluids are subject to future studies.

Acknowledgements

This work was funded by the German Science Foundation under grant number JA905 and SE1118 within the priority program "Nano- and Microfluidics" SPP 1164. We also acknowledge the generous support of Si wafers by Wacker, Burghausen, Germany.

Bibliography

J. Becker. *Numerische Simulation der Bildung fluider Strukturen auf inhomogenen Oberflächen.* Thesis, Bonn, 2004.

J. Becker, G. Grün, R. Seemann, H. Mantz, K. Jacobs, K. R. Mecke, and R. Blossey. Complex dewetting scenarios captured by thin-film models. *Nature Mat.*, 2:59–63, 2003.

J. Bischof. Dewetting modes of thin metallic films: Nucleation of holes and spinodal dewetting. Thesis, Konstanz, 1996.

J. Bischof, D. Scherer, S. Herminghaus, and P. Leiderer. Dewetting modes of thin metallic films: Nucleation of holes and spinodal dewetting. *Phys. Rev. Lett.*, 77:1536–1539, 1996.

R. Blossey. Nucleation at first-order wetting transitions. *Int. J. Mod. Phys. B*, 9:3489–3525, 1995.

F. Brochard-Wyart and J. Daillant. Drying of solids wetted by thin liquid films. *Can. J. Phys.*, 68:1084–1088, 1989.

S. G. Croll. Origin of residual internal-stress in solvent-cast thermoplastic coatings. *J. Appl. Polym. Sci.*, 23:847–858, 1979.

S. Dietrich. Wetting phenomena. volume 12 of *Phase Transitions and Critical Phenomena*, page 1. Academic Press, London, 1988.

R. Fetzer, K. Jacobs, A. Münch, B. Wagner, and T. P. Witelski. New slip regimes and the shape of dewetting thin liquid films. *Phys. Rev. Lett.*, 95:127801, 2005.

R. Fetzer, M. Rauscher, J. Becker, G. Grün, R. Seemann, K. Jacobs, and K Mecke. Thermal noise induences fuid fow in thin films during spinodal dewetting. *to be published*, 2006.

A. N. Frumkin. On the phenomena of wetting and sticking of bubbles (in Russian). *Zh. Fiz. Khim.*, 12:337, 1938.

S. Herminghaus. Polymer thin films and surfaces: Possible effects of capillary waves. *Eur. Phys. J. E*, 8:237–243, 2002.

S. Herminghaus, K. Jacobs, K. Mecke, J. Bischof, A. Fery, M. Ibn-Elhaj, and S. Schlagowski. Spinodal dewetting in liquid crystal and liquid metal films. *Science*, 282:916–919, 1998.

S. Herminghaus, K. Jacobs, and R. Seemann. The glass transition of thin polymer films: Some questions, and a possible answer. *Eur. Phys. J. E*, 5:531–538, 2001.

S. Herminghaus, K. Jacobs, and R. Seemann. Viscoelastic dynamics of polymer thin films and surfaces. *Eur. Phys. J. E*, 12:101–110, 2003.

S. Herminghaus, R. Seemann, and K. Jacobs. Generic morphologies of viscoelastic dewetting fronts. *Phys. Rev. Lett.*, 89:056101, 2002.

S. Herminghaus, R. Seemann, and K. Landfester. Polymer surface melting mediated by capillary waves. *Phys. Rev. Lett.*, 93:017801, 2004.

J. N. Israelachvili. *Intermolecular and Surface Forces.* Academic Press, London, 1992.

K. Jacobs, S. Herminghaus, and K. R. Mecke. Thin liquid polymer films rupture via defects. *Langmuir*, 14:965–969, 1998a.

K. Jacobs, R. Seemann, and K. Mecke. *Statistical Physics and Spatial Statistics*. Springer, Heidelberg, 2000.

K. Jacobs, R. Seemann, G. Schatz, and S. Herminghaus. Growth of holes in liquid films with partial slippage. *Langmuir*, 14:4961, 1998b.

K. Kargupta, R. Konnur, and A. Sharma. Spontaneous dewetting and ordered patterns in evaporating thin liquid films on homogeneous and heterogeneous substrates. *Langmuir*, 17:1294–1305, 2001.

H. I. Kim, C. M. Mate, K. A. Hannibal, and S. S. Perry. How disjoining pressure drives the dewetting of a polymer film on a silicon surface. *Phys. Rev. Lett.*, 82:3496–3499, 1999.

R. Konnur, K. Kargupta, and A. Sharma. Instability and morphology of thin liquid films on chemically heterogeneous substrates. *Phys. Rev. Lett.*, 84:931–934, 2000.

J. Koplik and J. R. Banavar. Molecular simulations of dewetting. *Phys. Rev. Lett.*, 84: 4401–4404, 2000.

K. Mecke. *Integralgeometrie in der Statistischen Physik - Perkolation, komplexe Flüssigkeiten und die Struktur des Universums, Reihe Physik Bd. 25*. Verlag Harri Deutsch, Frankfurt a.M., 1994.

V. S. Mitlin. Dewetting of solid surface: Analogy with spinodal decomposition. *J. Colloid Interface Sci.*, 156:491–497, 1993.

V. S. Mitlin. On dewetting conditions. *Colloid Surf. A-Physicochem. Eng. Asp.*, 89: 97–101, 1994.

A. Münch. Dewetting rates of thin liquid films. *J. Phys.-Condes. Matter*, 17:S309–S318, 2005.

C. Neto, K. Jacobs, R. Seemann, R. Blossey, J. Becker, and G. Grün. Correlated dewetting patterns in thin polystyrene films. *J. Phys.-Condes. Matter*, 15:S421–S426, 2003a.

C. Neto, K. Jacobs, R. Seemann, R. Blossey, J. Becker, and G. Grün. Satellite hole formation during dewetting: Experiment and simulation. *J. Phys.-Condes. Matter*, 15:3355–3366, 2003b.

D. Podzimek, A. Saier, R. Seemann, K. Jacobs, and S. Herminghaus. A universal nucleation mechanism for solvent cast polymer film rupturel. *arXiv:condmat/0105065*, 2001.

G. Reiter. Dewetting of thin polymer films. *Phys. Rev. Lett.*, 68:75–78, 1992.

G. Reiter, M. Hamieh, P. Damman, S. Sclavons, S. Gabriele, T. Vilmin, and E. Raphaël. Residual stresses in thin polymer films cause rupture and dominate early stages of dewetting. *Nat. Mater.*, 4:754–758, 2005.

E. Ruckenstein and R. K. Jain. Spontaneous rupture of thin liquid films. *J. Chem. Soc. Faraday Trans. II*, 70:132–147, 1974.

M. Schick. *Liquids at Interfaces*. Elsevier Science, Amsterdam, 1989.

R. Seemann, S. Herminghaus, and K. Jacobs. Dewetting patterns and molecular forces: A reconciliation. *Phys. Rev. Lett.*, 86:5534–5537, 2001a.

R. Seemann, S. Herminghaus, and K. Jacobs. Gaining control of pattern formation of dewetting liquid films. *J. Phys.-Cond. Mat.*, 13:4925–4938, 2001b.

R. Seemann, S. Herminghaus, and K. Jacobs. Shape of a liquid front upon dewetting. *Phys. Rev. Lett.*, 87:196101, 2001c.

R. Seemann, S. Herminghaus, C. Neto, S. Schlagowski, D. Podzimek, R. Konrad, H. Mantz, and K. Jacobs. Dynamics and structure formation in thin polymer melt films. *J. Phys.-Condes. Matter*, 17:S267–S290, 2005.

R. Seemann, K. Jacobs, and R. Blossey. Polystyrene nanodroplets. *J. Phys.-Condes. Matter*, 13:4915–4923, 2001d.

R. Seemann, K. Jacobs, K. Landfester, and S. Herminghaus. submitted. *J. Pol. Sci. B*, 2006.

M. Sferrazza, M. Heppenstall-Butler, R. Cubitt, D. Bucknall, J. Webster, and R. A. L. Jones. Interfacial instability driven by dispersive forces: The early stages of spinodal dewetting of a thin polymer film on a polymer substrate. *Phys. Rev. Lett.*, 81:5173–5176, 1998.

A. Sharma. Many paths to dewetting of thin films: anatomy and physiology of surface instability. *Eur. Phys. J. E*, 12:397–407, 2003.

A. Sharma and R. Khanna. Pattern formation in unstable thin liquid films. *Phys. Rev. Lett.*, 81:3463–3466, 1998.

A. Sharma and R. Khanna. Pattern formation in unstable thin liquid films under the influence of antagonistic short- and long-range forces. *J. Chem. Phys.*, 110:4929–4936, 1999.

T. G. Stange, D. F. Evans, and W. A. Hendrickson. Nucleation and growth of defects leading to dewetting of thin polymer films. *Langmuir*, 13:4459–4465, 1997.

M. Sze. *Physics of Semiconductor Devices*. Wiley, New York, 1981.

U. Thiele. Open questions and promising new fields in dewetting. *Eur. Phys. J. E*, 12: 409–416, 2003.

U. Thiele, K. Neuffer, Y. Pomeau, and M. G. Velarde. On the importance of nucleation solutions for the rupture of thin liquid films. *Colloid Surf. A*, 206:135–155, 2002.

U. Thiele, M. G. Velarde, and K. Neuffer. Dewetting: Film rupture by nucleation in the spinodal regime. *Phys. Rev. Lett.*, 87:016104, 2001.

O. K. C. Tsui, Y. J. Wang, H. Zhao, and B. Du. Some views about the controversial dewetting morphology of polystyrene films. *Eur. Phys. J. E*, 12:417–423, 2003.

A. Vrij. Possible mechanism for the spontaneous rupture of thin free liquid films. *Disc. Faraday Soc.*, 42:23–33, 1966.

M. B. Williams and S. H. Davis. Nonlinear theory of film rupture. *J. Colloid Interface Sci.*, 90:220–228, 1982.

R. Xie, A. Karim, J. F. Douglas, C. C. Han, and R. A. Weiss. Spinodal dewetting of thin polymer films. *Phys. Rev. Lett.*, 81:1251–1254, 1998.

Structure Formation in Thin Liquid Films

Uwe Thiele

Max-Planck-Institut für Physik komplexer Systeme, Dresden, Germany

Abstract We outline some recent developments in the theoretical description of structure formation in thin liquid films. The main focus is systems involving a single layer of liquid on a solid substrate that can be described using an evolution equation for the film thickness profile. We review the history of the subject and we sketch important experimental and theoretical results and practical applications. After discussing the classification of the different cases, we introduce the common mathematical framework for studies of thin films of soft matter, namely by deriving the generic evolution equation for such films from the Navier-Stokes equations. In the main part we first introduce the different possible geometries and the transitions between them, i.e. from homogeneous to inhomogeneous substrates, or from horizontal to inclined substrates. We then present the physical questions posed by the individual systems and discuss approaches and results for:

- Dewetting on a horizontal homogeneous substrate. We investigate the solution structure and its consequences for the system behavior. For the initial film rupture we distinguish nucleation-dominated and instability-dominated behavior for linearly unstable thin films.
- Dewetting on a horizontal inhomogeneous substrate. The solution structure of the governing equation is analysed as a function of the strength of a chemical heterogeneity. We describe a pinning-coarsening transition with a large range of multistability, implying a large hysteresis and strong dependence on initial conditions and noise.
- Heated thin films on a horizontal homogeneous substrate. We discuss nucleation and drop solutions and show that it is possible to construct all drop solutions separated by dry regions. Incorporating a disjoining pressure allows to study the coarsening behaviour of the drop pattern.
- Sliding drops on an inclined homogeneous substrate. Using a model that incorporates a disjoining pressure allows to calculate the frequently used ad-hoc parameters of models for moving contact lines from surface chemistry. The involved transition from a Cahn-Hilliard-like to a Kuramoto-Sivashinsky-like dynamics that occurs for increasing inclination angle is analyzed for heated films.
- Transverse instabilities of a liquid ridge are discussed encompassing all the above geometries. Particular attention is given on the stabilization of such an instability due to stripe-like heterogeneities for a resting ridge on a horizontal substrate and on the drastic change in the mode type when inclining the substrate. It changes from a symmetric varicose mode (horizontal substrate) via an asymmetric varicose mode and via an asymmetric zigzag mode to decoupled front and back modes.

Finally, we shortly discuss extensions of thin film studies beyond the case of a single evolution equation. In particular, we introduce two different models based on two coupled evolution equations describing the dynamics of dewetting of a two-layer thin film and the chemically driven self-propelled movement of droplets, respectively.

1 Introduction

1.1 History

For centuries thin liquid films have attracted the interest of scientists and laymen alike due to their fascinating behavior and presence in everyday life. The most prominent example is probably soap films whose beauty is appreciated by almost everyone. Their properties were investigated by scientists like Newton (1730a), Thomson (1887) or Plateau (1873).

Besides these freely suspended films, thin liquid layers between two solid substrates were also studied very early. Films and drops between two parallel plates were used already by Newton (1730b) to investigate capillary effects (following the experimentalist Hauksbee (1708, 1710)). However, our understanding is rather based on the explanations given later by Young (1805) and Laplace (1806). The physical principles of the use of thin films for lubrication purposes were analyzed by Reynolds (1886). He laid the foundation for their description as slow viscous flow by developing the 'lubrication approximation' that was later built on by Sommerfeld (1904) and others.

An intermediate case between free films and films between two solid supports are liquid films that are bounded on one side by a solid substrate but have a free surface on the other side. This type of films will concern us here. They were mentioned early in connection with liquid flow driven by surface tension gradients by Thomson (1855) in his description of the phenomenon of the tears of wine, and also by Tomlinson (1870) and Marangoni (1871) referring to experiments on the spreading of surface active substances on thin films of water[1]. Free surface thin films are also the basis for studying surface waves and localized structures at the surface of a thin viscous layer flowing down an inclined plate. Starting with the experiments by the Kapitzas (Kapitza, 1949; Kapitza and Kapitza, 1949) the system became paradigmatic for the study of this type of structure formation (see the review by Chang (1994) and the beginning of Section 5.2).

Nowadays, thin liquid films on solid substrates are studied in a wide range of fields and have numerous applications. The spectrum ranges from films of sub-micrometer thickness (Ruckenstein and Jain, 1974; Kheshgi and Scriven, 1991) to studies of lava flows (Huppert, 1982; Balmforth et al., 2004). This indicates that 'thin' does not refer to an absolute film thickness, but rather to the fact that the films are thin as compared to typical length-scales parallel to the substrate. Representative examples of medical interest are the tear film in the eye (Lin and Brenner, 1982; Sharma and Ruckenstein, 1985) and the aqueous lung lining (Gaver and Grotberg, 1990). Heat and mass transfer devices, like falling film evaporators rely on the heat-transfer properties and stability of falling thin liquid films (Bankoff, 1994). Thin films also gained increasing importance

---

[1]The history of the study of capillary phenomena (Hardy, 1922; Millington, 1945; Scriven and Sternling, 1960) offers more examples involving thin films.

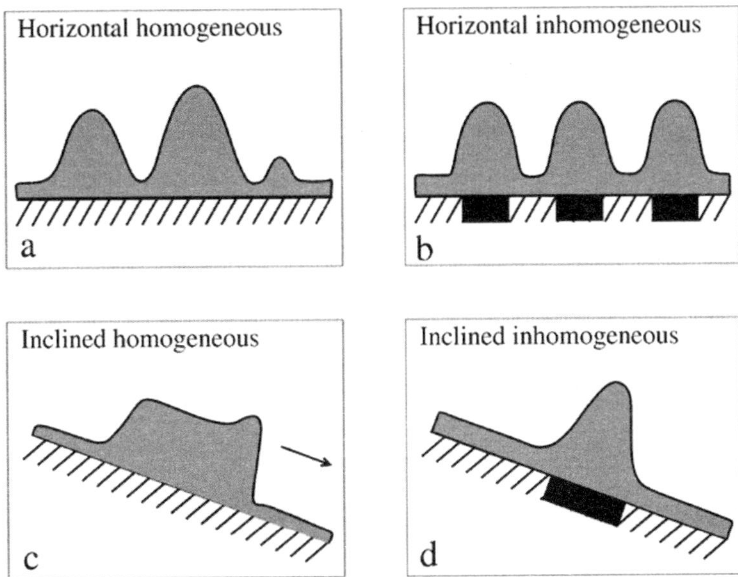

Figure 1. Sketch of the four basic geometries involving a single liquid layer on a solid substrate.

in coating technology. In addition to the use as protective and aesthetic coatings requiring stable homogeneous films, the capacity of the films to structure themselves in the nanometer range renewed the interest in their dynamics. The importance of structured and non-structured coatings in many fields of modern technology has led to a still growing series of studies of the structuring process of ultrathin films with thicknesses below 100 nm that are unstable due to effective molecular forces. These experiments on dewetting were pioneered by Reiter (1992) using polymer films and are continued by a number of groups (for a review see Thiele (2003a) and the chapter by Seemann et al. in the present book). Thereby all phases of the process are of interest (Brochard-Wyart and Daillant, 1989; Brochard-Wyart et al., 1992): the initial film rupture (Reiter, 1992; Seemann et al., 2001a), the growth of individual holes in the film (Redon et al., 1991; Seemann et al., 2001b), the evolution of the resulting hole pattern (Sharma and Reiter, 1996), and the stability of the individual dewetting fronts (Brochard-Wyart and Redon, 1992; Reiter and Sharma, 2001). The cursory overview of the literature given at this point is supplemented by more detailed synopses in the presentations of the specific physical situations below.

1.2 Classification

In Sections 1 and 2 we introduce the general subject of thin liquid films whereas the following Sections are devoted to more detailed presentations of the specific models used and the individual physical situations studied. We do not discuss every situation individually, but group them according to their symmetry properties to emphasize that these are primarily responsible for the qualitative behavior of thin films. Based on substrate properties and acting forces one can distinguish four basic geometries as sketched in Figure 1:

(a) A film on a horizontal homogeneous substrate is in a situation of maximal symmetry. The situation is invariant with respect to translation along the substrate and reflection at a plane orthogonal to the substrate[2]. This type of films will be introduced in Section 3 using the examples of dewetting on a smooth homogeneous substrate (Reiter, 1992) and of the pattern formation of a thin film on a homogeneously heated substrate (VanHook et al., 1997).

There are three ways to break these symmetries to arrive at the less symmetric situations (b) to (d):

(b) The translational symmetry is broken for a film on a horizontal inhomogeneous substrate as encountered for dewetting on a physically or chemically patterned substrate (Rehse et al., 2001) or for a film on an inhomogeneously heated plate (Burelbach et al., 1990; Tan et al., 1990). The heterogeneity especially influences processes where translational modes are involved as is the case with coarsening. This case is introduced in Section 4.

(c) The reflection symmetry is broken if an additional force is acting parallel to the substrate as for a flowing film or sliding droplet on an inclined homogeneous substrate (Kapitza and Kapitza, 1949; Podgorski et al., 2001) where the driving force is gravity. Also a temperature or chemical gradient along the substrate leads to this situation (Brochard, 1989). Here we will use the inclined case as example to discuss moving drops and unstable fronts. It is introduced in Section 5.

(d) Both, the translational and reflection symmetry are broken on an inclined inhomogeneous substrate as encountered for films flowing on an inclined locally heated plate (Kabov and Marchuk, 1998) or droplets pinned by heterogeneities on an incline (Quéré et al., 1998).

All the situations covered by (a) to (d) can be described by a single partial differential equation for the evolution of the film thickness profile. This allows to move freely between the different geometries and to draw on results from the respective 'neighboring' ones. Excluding evaporation and condensation processes the governing equation is of the most general type for the evolution of a single conserved order parameter field (Langer, 1992). The basic formalism is introduced in Section 2, whereas Sections 3 to 5 are devoted to the above introduced situations (a) to (c). Situation (d) will only be mentioned briefly at the end of Section 5. Section 6 is devoted to the question of transverse contact line instabilities that are important in all the situations (a) to (d).

Finally, in Section 7 we abandon the framework of a single evolution equation by focusing our interest on situations involving thin films where more degrees of freedom have to be taken into account. In the simplest case, two order parameter fields instead of only one are needed to model the dynamics. On general grounds one can distinguish

---

[2]Note, that the symmetry does not refer to the thickness profile of the film itself because it may evolve towards less symmetric states.

two situations: (i) the second field besides the film thickness profile represents also a conserved order parameter like for a two-layer ultrathin film studied in Section 7.1; (ii) the second field corresponds to a non-conserved order parameter like for the chemically driven droplets discussed in Section 7.2. In Section 8 we draw some conclusions, discuss open questions and give an outlook.

1.3 Theoretical Approaches

A well established approach to determine the shape and stability of static fronts, droplets and ridges on homogeneous or heterogeneous substrates is based on variations of a suitable free energy functional (de Gennes, 1985; Sekimoto et al., 1987; Lenz and Lipowsky, 1998; Bauer et al., 1999; Bauer and Dietrich, 2000; Brinkmann and Lipowsky, 2002). It has the advantage that it is not restricted to small equilibrium contact angles. However, it is not suited to describe dynamic phenomena or 'dynamic aspects' of the static problem, as for instance, the most dangerous mode for the Rayleigh-like instability of a liquid ridge on a horizontal substrate (cp. results of Sekimoto et al. (1987) and Thiele and Knobloch (2003)).

The theoretical description of the dynamics of thin films with a free surface goes back half a century. Besides modeling the full Navier-Stokes (Salamon et al., 1994; Krishnamoorthy et al., 1995) or Stokes (Boos and Thess, 1999) equations with moving boundaries one can derive a reduced model based on the 'long-wave approximation' (Oron et al., 1997). The latter can be applied to all the structuring processes covered here, because the films are thin as compared to the lateral extension of the evolving long-wave structures. The occurrence of short-wave structures whose lateral extension is of the order of the film thickness as convection cells in films heated from below – as described by Bénard (1900) and Rayleigh (1916) – is avoided by choosing parameters properly. Also the interaction of short-wave and long-wave structures is not examined here (but see Golovin et al. (1994); VanHook et al. (1997)).

The long-wave approximation was used to derive dynamical equations for the evolution of the film thickness profile for falling films by Benney (1966), for free films by Vrij (1966)[3], for films on a heated horizontal substrate by Burelbach et al. (1988), for films on heated slightly inclined plates by Oron and Rosenau (1992), and for ultrathin films on a horizontal substrate by Ruckenstein and Jain (1974). In the latter case the films may be unstable due to acting effective molecular interactions that are incorporated in the form of an additional pressure term into the governing equations. This so-called 'disjoining pressure' was introduced by Derjaguin and coworkers in connection with work on the forces acting between two solid plates separated by a thin film (Derjaguin et al., 1987; Dzyaloshinskii et al., 1960). In the simplest case the disjoining pressure results only from the apolar London–van der Waals dispersion forces (Ruckenstein and Jain, 1974), but there may be additional polar short-range contributions (Sharma, 1993b). Mitlin (1993) established the analogy between the surface instability observed for ultrathin films, called spinodal dewetting, and spinodal decomposition studied by Cahn and Hilliard (1958)[4].

---

[3]Note, that there only the linear stage of the evolution is covered.

[4]However, already Vrij (1966) noted the formal equivalence of the equation for free films to the equation of Cahn (1965) for concentration fluctuations in solutions.

Consequently, most results obtained for the decomposition of a binary mixture have a counterpart in the evolution of thin films on horizontal substrates and *vice versa*. One can generally say, that thin films on horizontal substrates follow a Cahn-Hilliard-like dynamics.

By inclining the substrate strongly, inertia has to be taken into account leading in the simplest case to the Benney equation (Benney, 1966). This equation was extended to include the effect of a heated substrate by Joo et al. (1991). Such Benney-like equations were used to perform linear stability analysis (Benjamin, 1957; Yih, 1963), weakly non-linear analysis (Benney, 1966; Gjevik, 1970; Shkadov, 1967), fully non-linear analysis (Joo et al., 1991), to study sideband instabilities (Lin, 1974) and solitary waves (Pumir et al., 1983). The small amplitude limit of the falling film equations corresponds to the Kuramoto-Sivashinsky equation (Kuramoto and Tsuzuki, 1976; Sivashinsky, 1977). Here, we will discuss Benney-type equations only briefly, but will demonstrate in Section 5.3 that for slightly inclined plates, where no inertia is included in the formulation, Kuramoto-Sivashinsky-type dynamics can be found.

2 The Derivation of the Film Thickness Evolution Equation

2.1 General Approach

A layer of liquid on a solid substrate with a free surface is called a *thin film* if its thickness is small compared to all relevant length scales parallel to the substrate. This refers to substrate properties like length scales of a heterogeneity as well as to typical lateral extensions of the surface profile of the film itself. The latter implies that the slope of the profile has to be small, but does not restrict the amplitude of its modulations. As a consequence, *thin film* does by no means refer to some absolute thickness measure but has to be defined individually for each physical situation studied.

In principle, all situations pictured in Figure 1 are governed by the Navier-Stokes equations with adequate boundary conditions at the substrate and the free surface, i.e. the liquid-gas interface. Usually, at the liquid-gas interface the balance of the stress-tensors is applied by assuming the gas to be passive. At the substrate usually the no-slip condition is applied. However, also a variety of slip conditions, as for instance, the Navier slip condition are also used (Oron et al., 1997; Münch, 2005). For the thin film geometry the velocities parallel to the substrate are large compared to the ones orthogonal to the substrate. Continuity then implies that the gradients orthogonal to the substrate are large compared to the ones parallel to the substrate. This allows to simplify the governing equations using the 'long-wave' or 'lubrication approximation' (Oron et al., 1997). A kinematic boundary condition for the free surface assures that the material boundary moves with the velocity of the liquid at the boundary. Putting all this together one derives a fourth order nonlinear partial differential equation describing the time evolution of the film thickness profile. For simplicity here we restrict ourselves to a physically two-dimensional situation where the film thickness $h(t, x)$ depends on the coordinate x only [5].

---

[5]Note, that in the literature there exist two different ways to count the dimensions. On the one hand, one can derive the dimension from the physical situation, i.e. a drop on a plate represents

2.2 Basic Equations

The derivation of the thin film equation starts with the transport equation for the momentum density (Navier-Stokes equations)

$$\varrho \frac{d\vec{v}}{dt} = \nabla \cdot \underline{\tau} + \vec{f} \tag{2.1}$$

where $\vec{v}(x,z)$ and $\vec{f}(x,z)$ are the velocity and a body force field, respectively. We use

$$\vec{v} = \begin{pmatrix} u \\ w \end{pmatrix}, \qquad \vec{f} = \begin{pmatrix} f_1 \\ f_2 \end{pmatrix} \qquad \text{and} \qquad \nabla = \begin{pmatrix} \partial_x \\ \partial_z \end{pmatrix}.$$

In the following we will denote partial derivatives with respect to i either by ∂_i or simply by the subscript i. The body force may be potential, i.e. given by $\vec{f} = -\nabla\phi$. The stress tensor writes

$$\underline{\tau} = -p\underline{I} + \eta(\nabla\vec{v} + (\nabla\vec{v})^T) \tag{2.2}$$

where $p(x,z)$ stands for the pressure field and $\underline{I}$ is the unity tensor. The material derivative is defined by

$$\frac{d}{dt} = \frac{\partial}{\partial t} + (\vec{v} \cdot \nabla). \tag{2.3}$$

The transport equation for the energy writes

$$\varrho c_p \frac{dT}{dt} = k_{th}\nabla^2 T + \frac{\eta}{2}\left[\nabla\vec{v} + (\nabla\vec{v})^T\right]^2 \tag{2.4}$$

where $T(x,z)$ is the temperature field. The last term corresponds to the mechanical work due to inner tensions $(\underline{\tau} \cdot \nabla) \cdot \vec{v}$. Due to its smallness it will be neglected in the following. Finally, for an incompressible liquid one has the continuity equation

$$\nabla \cdot \vec{v} = 0. \tag{2.5}$$

The parameters ρ, η, c_p, and k_{th} are the density, dynamic viscosity, specific heat and thermal conductivity of the liquid, respectively. The kinematic viscosity is $\nu = \eta/\varrho$ whereas the thermal diffusivity is $\kappa_{th} = k_{th}/\varrho c_p$.

2.3 Boundary Conditions

The transport equations are accompanied by boundary conditions at the smooth solid substrate and the free surface. We assume for the velocity field at the substrate ($z = 0$) the no-slip and the no-penetration condition

$$\vec{v} = 0. \tag{2.6}$$

---

a three-dimensional situation and an orthogonal cut through a liquid ridge represents a two-dimensional situation. This is what we use throughout the present work. On the other hand, one can base the dimension count on the spatial dimensions the film thickness profile depends on, i.e. drop and cut through a ridge are represented by two- and one-dimensional profiles, respectively.

At the free surface $[z = h(t,x)]$ one has the kinematic condition (surface follows flow field)

$$w = \partial_t h + u \partial_x h \tag{2.7}$$

and the force equilibrium

$$(\underline{\tau} - \underline{\tau}_{air}) \cdot \vec{n} = K\gamma \vec{n} + (\partial_s \gamma)\vec{t} \tag{2.8}$$

where the surface derivative is defined by $\partial_s = \vec{t} \cdot \nabla$ and we assume that the ambient air does not transmit any force ($\underline{\tau}_{air} = 0$). The term $p_L = -\gamma K = \gamma \nabla \cdot \vec{n}$ corresponds to the Laplace or curvature pressure whereas $\partial_s \gamma$ results from the variation of the surface tension along the surface (caused, for instance, by solutal or thermal Marangoni effects). The latter is modeled in the simplest case by a linear dependence of the surface tension on temperature, $\gamma = \gamma_0 + \gamma_T(T_0 - T)$, where γ_0 is the surface tension at the reference temperature T_0 and $\gamma_T = d\gamma/dT$ at γ_0.

$$\vec{n} = \frac{(-\partial_x h, 1)}{\left(1 + (\partial_x h)^2\right)^{1/2}}, \qquad \vec{t} = \frac{(1, \partial_x h)}{\left(1 + (\partial_x h)^2\right)^{1/2}}, \qquad K = \frac{\partial_{xx} h}{\left(1 + (\partial_x h)^2\right)^{3/2}}$$

are the normal vector, tangent vector and curvature of the surface, respectively. The boundary condition (2.8) is of vectorial character, i.e. one can derive two scalar conditions by projecting it onto $\vec{n}$ and $\vec{t}$, respectively.

$$\vec{t} : \quad \eta\left[(u_z + w_x)(1 - h_x^2) + 2(w_z - u_x)h_x\right] = \partial_s \gamma (1 + h_x^2) \tag{2.9}$$

$$\vec{n} : \quad p + \frac{2\eta}{1 + h_x^2}\left[-u_x h_x^2 - w_z + h_x(u_z + w_x)\right] = -\frac{\gamma h_{xx}}{(1 + h_x^2)^{3/2}} \tag{2.10}$$

For the temperature field we will assume a constant temperature at the substrate and Newton's cooling law at the free surface, i.e.

$$T = T_0 \qquad \text{at} \qquad z = 0$$

$$\text{and} \qquad \kappa \vec{n} \cdot \nabla T + \alpha_{th}(T - T_\infty) = 0 \qquad \text{at} \qquad z = h(x),$$

respectively. The temperature of the ambient gas is T_∞ and α_{th} is the heat transfer coefficient.

2.4 Incorporation of Interaction With the Substrate

Interestingly, the above transport equations and boundary conditions of classical hydrodynamics are not sufficient to account for all situations involving thin films or drops. For a static droplet on a solid substrate (Figure 2 (a)) in an isothermal situation, we have $\vec{v} = 0$ everywhere and equation (2.1) reduces to the static equation

$$-\nabla p + \vec{f} = 0. \tag{2.11}$$

Assuming an isothermal situation and neglecting hydrostatic effects and other body forces ($\vec{f} = 0$), the Laplace pressure (equation 2.10)) determines the solution of (2.11) as

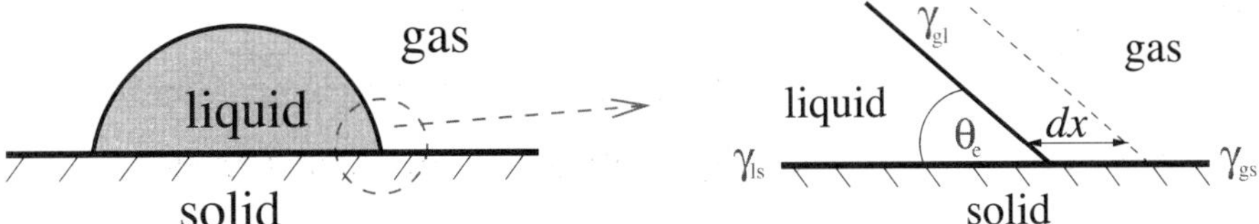

Figure 2. Sketch of a cap-like droplet and the geometry of a static three-phase contact line, indicating gas-solid (γ_{gs}), gas-liquid (γ_{gl}), and liquid-solid (γ_{ls}) interfacial tensions and the equilibrium contact angle θ_e.

a spherical cap-like droplet (or a flat film); see Figure 2. However, even for a fixed volume the radius of the droplet is still a free parameter. One needs an additional condition at the three-phase contact line. Such a condition can be derived from the involved interface energies using the invariance of the total energy with respect to translation (de Gennes, 1985). It is the well known Young-Laplace law

$$\gamma_{lg} \cos\theta_e = \gamma_{gs} - \gamma_{sl} \tag{2.12}$$

where θ_e is the equilibrium contact angle and the γ_{ij} are the interfacial tensions (see Figure 2 (b)).

The Young-Laplace law describes well situations of so-called 'partial wetting' where $0 < \theta_e < \pi$, i.e. where $-1 < (\gamma_{gs} - \gamma_{sl})/\gamma_{lg} < 1$. Corresponding droplets are depicted in Figure 3 (b). To cover the remaining cases also it is practical to introduce the spreading coefficient (de Gennes, 1985)

$$S = \gamma_{gs} - \gamma_{lg} - \gamma_{sl}, \tag{2.13}$$

that measures the energy difference of a dry substrate and a substrate with a liquid film. For partial wetting the combination with equation (2.12) yields

$$S = \gamma_{lg} \left(\cos \theta_e - 1\right) \tag{2.14}$$

implying $\theta_e = \sqrt{-2S/\gamma}$ for small equilibrium contact angles, i.e. in long-wave approximation.

The spreading coefficient also allows to quantify in a differentiated manner the non-wetting situation (Figure 3 (a)) where always $\theta_e = \pi$ but S is only restricted by $S/\gamma_{lg} \leq -2$. The same is valid for the situation of complete wetting (Figure 3 (c)) where always $\theta_e = 0$ (i.e. the liquid forms a film) but S is only restricted by $S \geq 0$. In this way static situations are well described.

However, a contact line still causes serious problems in hydrodynamics, especially in situations where it moves like for a droplet that slides down an incline. If the drop surface truly touches the substrate at the contact line and one does not relax the classical no-slip condition (equation (2.6)) then the viscous dissipation diverges at the contact line and it can not move (de Gennes, 1985).

A second problem concerns ultrathin films of thicknesses below 100 nm. Normally, the energy of an interface between two substances, i.e. the interface tension, is calculated

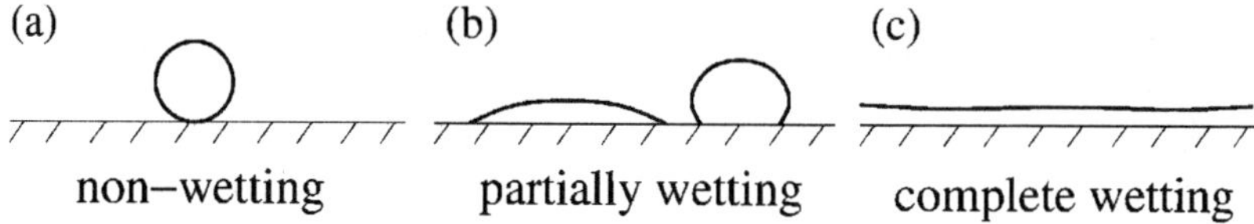

Figure 3. Sketches of the three qualitatively different wetting situations for a simple liquid on a smooth solid substrate: (a) non-wetting, (b) partially wetting, and (c) complete wetting.

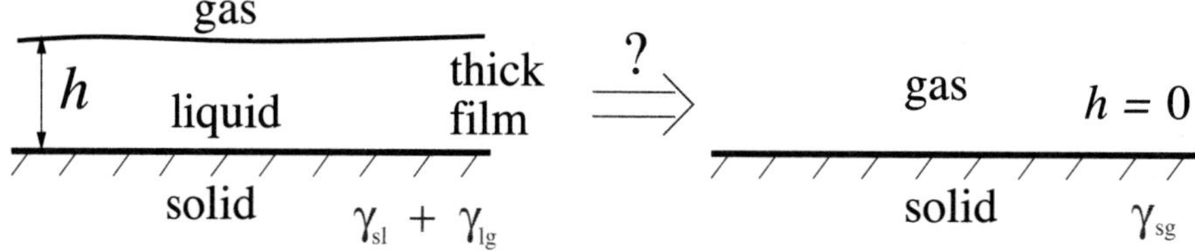

Figure 4. Sketch of the transition from a 'bulk film' to a dry substrate.

assuming both are bulk materials. This is still a very good approximation for a thick film depicted in Figure 4 (a). There the overall interfacial energy is the sum of the solid-liquid one (γ_{sl}) and the liquid-gas one (γ_{lg}). Without the film (Figure 4 (b)) one has only the energy of the solid-gas interface (γ_{sg}). However, the question arises how does one model the transition from (a) to (b), i.e. what are the surface energies for situations where the film can not be regarded as a bulk volume.

Derjaguin and coworkers found that for ultrathin films there is an additional energy $V(h)$ depending on the thickness of the film (Dzyaloshinskii et al., 1960; Israelachvili, 1992). The thickness dependence leads to an additional attractive or repulsive force between the two film interfaces. This force is normally included into the hydrodynamic formalism as an additional pressure term $\Pi = -dV/dh$, the so-called 'disjoining or conjoining pressure' (de Gennes, 1985; Teletzke et al., 1988; Oron et al., 1997). Is it introduced either in the normal force boundary condition supplementing the Laplace pressure

$$p_L \to p_L - \Pi(h)$$

or as an additional body force in the Navier Stokes equations (see de Gennes (1985)):

$$\vec{f}_{add} = -\nabla \phi_{add} \qquad \text{with} \qquad \phi_{add} = \Pi(z) - \Pi(h)$$

Both ways lead to the same result. Here we follow the first one.

The disjoining pressure can be calculated for specific intermolecular interactions. Assuming a long-range apolar van der Waals interaction one finds

$$\Pi_{vdW}(h) = \frac{2S_a d_0^2}{h^3} \tag{2.15}$$

where S_a is the apolar contribution to the spreading coefficient. The length d_0 is often introduced as a molecular cut-off the so-called Born repulsion length, $d_{Born} = 0.158\,\mathrm{nm}$ (Sharma, 1993b,a). $A = -12\pi S_a d_{Born}^2$ is the Hamaker constant that can be calculated from the optical indices of the involved materials (Israelachvili, 1992). For a short-range polar interaction (like, for instance, arising from the interaction of the electric double layers in a thin film of an electrolyte) one finds (Israelachvili, 1992; Probstein, 1994)

$$\Pi_p(h) = \frac{S_p}{l} e^{(d_0 - h)/l} \tag{2.16}$$

where S_p is the polar contribution to the spreading coefficient and l is the correlation length of the polar interaction. One example of a commonly used expression for the disjoining pressure is the combination of the above long-range and short-range parts which in dimensionless form writes (Sharma, 1993b; Thiele et al., 2001a)

$$\Pi(h) = \frac{b}{h^3} - e^{-h} \tag{2.17}$$

where we introduced the ratio of apolar and polar interaction $b = 2S_a d_0^2/|S_p|l^2 e^{d_0/l}$ and assumed $S_a > 0$ and $S_p < 0$ (h is in units of l). Figure 5 gives its dependence on film thickness for different values of b.

The choice of constants in equations (2.15) and (2.16) ensures that the total surface energy for a thin film $\gamma_{ls} + \gamma_{gl} + V(h)$ [where $V(h) = -\int \Pi(h)dh$] correctly interpolates between the thick film and the dry substrate, i.e. between the two situations depicted in Figure 4. For thick films $h \to \infty$ and $V(h) \to 0$, the energy is $\gamma_{ls} + \gamma_{gl}$ as expected. At the cut-off height $h = d_0$ one has $V(d_0) = S_a + S_p = S$, i.e. taking into account equation (2.13) the surface energy is just γ_{gs}. Using equation (2.14) small contact angles are given by $\theta_e = \sqrt{-2V(d_0)/\gamma}$.

If the part of the interaction with the shortest range is stabilizing, a very thin precursor film of thickness h_p can be found on the macroscopically 'dry' parts of the substrate. The precursor film allows the contact line to move (see sketch in Figure 6). However, the three-phase contact line must now be seen as a contact region. Note, that with the disjoining pressure the precursor thickness is not an ad-hoc parameter but already a result of the model. If there exists no other equilibrium thickness besides h_p, h_p is given by $\Pi(h_p) = 0$. Otherwise it can be obtained via a Maxwell construction (Mitlin, 1993; Thiele et al., 2001b).

As we will see later in section 3.1, in the former case the equilibrium contact angle can be identified with $\sqrt{-2V(h_p)/\gamma_{lg}}$. Consistency with the above expression implies $d_0 = h_p$. Then θ_e corresponds to the contact angle for the 'moist' case of de Gennes (1985). Keeping $d_0 = d_{Born} < h_p$, however, leads to the introduction of two different equilibrium contact angles: one for a truly dry substrate and one for the moist one (compare with de Gennes (1985)).

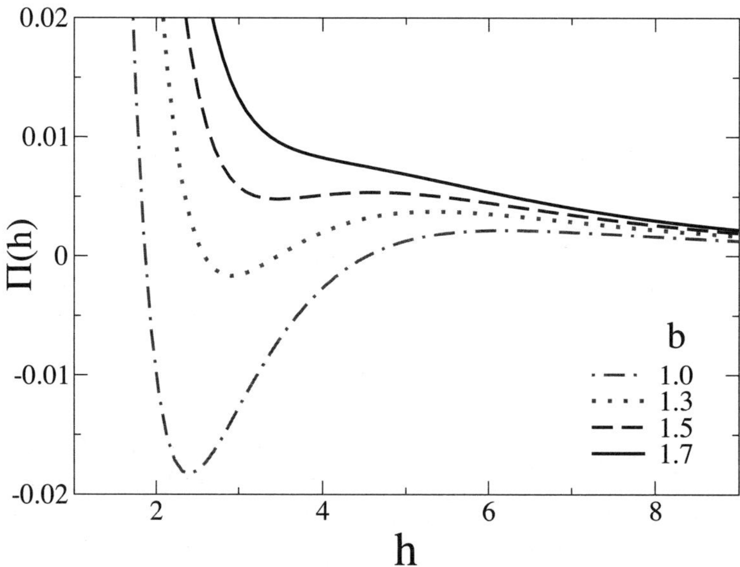

Figure 5. The disjoining pressure given by equation (2.17) for different values of the ratio of apolar and polar interactions b.

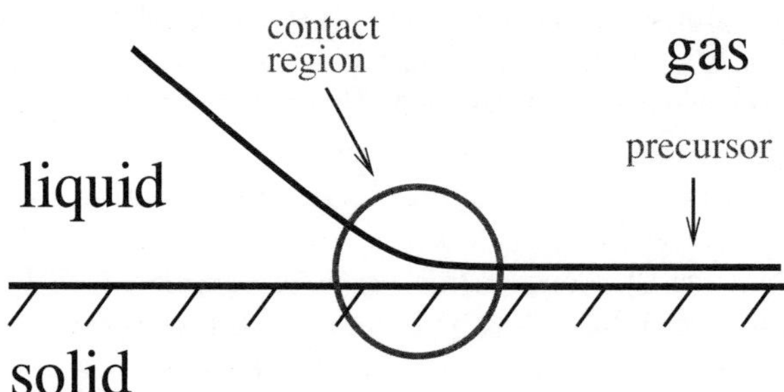

Figure 6. Sketch of a three-phase contact line with a precursor film on the 'dry' substrate.

Note finally, that for the disjoining pressure used here [Equations (2.15) and (2.16)] the choice $S_a > 0$ and $S_p < 0$ is usually described as 'a combination of a stabilizing long-range van der Waals and a destabilizing short-range polar interaction'. However, this does not reflect the complete picture due to a subtle feature of the combination of the exponential and the power law. Combining a term $\sim 1/h^3$ and one $\sim \exp(-h)$ leads for some choice of parameters to a dominance of $1/h^3$ for large and very small h. The exponential term only dominates for intermediate thicknesses. This implies that the above description only covers part of the feature of the disjoining pressure and should be used with caution.

2.5 Dimensionless Groups (Numbers)

General As a next step we introduce suitable scales for space, time, pressure, velocity and temperature and write the governing equations and boundary conditions in a dimensionless form. To find a suitable scaling is by no means a 'mechanical' task and requires the determination of a 'good' minimal set of parameters. At first one introduces a general rather abstract set of scales as follows:

$$
\begin{array}{ccc}
\text{dimensionless} & \text{Scale} & \text{dimensional} \\
t' & t_0 = l/U_0 & t = t_0 t' \\
x',\, z' & l & x = lx',\ z = lz' \\
\vec{v}' & U_0 & \vec{v} = U_0\vec{v}' \\
p' & P_0 = \varrho U_0^2 & p = P_0 p' \\
T' & \Delta T = T_0 - T_\infty & T = \Delta T T' + T_\infty
\end{array}
\qquad (2.18)
$$

The length l may be the mean film thickness, the precursor film thickness or another relevant vertical length scale. Introducing these scales, for instance, in the x-component of the momentum transport equations (2.1) and assuming $\vec{f} = \varrho\vec{g} = \varrho g\,(\sin\alpha, \cos\alpha)$ one obtains for a film on an incline

$$
\varrho\frac{U_0^2}{l}(u'_{t'} + u'u'_x + w'u'_{z'}) = -\frac{\varrho U_0^2}{l}p'_{x'} + \eta\frac{U_0}{l^2}(u'_{x'x'} + u'_{z'z'}) + \varrho g\sin\alpha \qquad (2.19)
$$

where α is the inclination angle. For simplicity we drop the dashes and introduce the dimensionless numbers

$$
\begin{array}{lll}
\text{Reynolds number} & \mathrm{Re} = \dfrac{U_0 l \varrho}{\eta} & \\[2ex]
\text{Froude number} & \mathrm{Fr} = \dfrac{U_0^2}{lg} &
\end{array}
\qquad (2.20)
$$

and get (now using both components of equation (2.1))

$$
\begin{aligned}
u_t + uu_x + wu_z &= -p_x + \frac{1}{\mathrm{Re}}(u_{xx} + u_{zz}) + \frac{\sin\alpha}{\mathrm{Fr}} \\
w_t + uw_x + ww_z &= -p_z + \frac{1}{\mathrm{Re}}(w_{xx} + w_{zz}) - \frac{\cos\alpha}{\mathrm{Fr}}
\end{aligned}
\qquad (2.21)
$$

The Reynolds number corresponds to the ratio of the selected velocity scale and the viscous one, whereas the Froude number corresponds to the squared ratio of the selected velocity scale to the gravitational one. The general scaling says nothing about the 'driving force' or the dominant time scale. The scaling becomes specific if the length l and velocity U_0 are identified. Note, that all scales in (2.20) are derived from l and U_0. In general this is not required. For a film flowing on an incline we illustrate this point by presenting three scalings based on different choices for the velocity scale U_0. They lead to specific expressions for the dimensionless numbers. These are, first, the viscous scaling

$$U_0 = \frac{\eta}{\varrho l}, \qquad \frac{1}{\mathrm{Re}} \to 1, \qquad \frac{1}{\mathrm{Fr}} \to \frac{gl^3\varrho^2}{\eta^2} =: G \qquad (2.22)$$

where G is the Galilei or Gravitation number; second, the falling film scaling

$$U_0 = \frac{\varrho g l^2 \sin\alpha}{\eta}, \qquad \frac{1}{\mathrm{Re}} \to \frac{1}{G\sin\alpha}, \qquad \frac{1}{\mathrm{Fr}} \to \frac{1}{G\sin\alpha}; \qquad (2.23)$$

and third, the surface tension scaling

$$U_0 = \frac{\gamma}{\eta}, \qquad \frac{1}{\mathrm{Re}} \to \frac{\eta^2}{\varrho l \gamma} = \frac{1}{S}, \qquad \frac{1}{\mathrm{Fr}} \to \frac{lg\eta^2\sin\alpha}{\gamma^2} = \frac{G\sin\alpha}{S^2} \qquad (2.24)$$

where S is the dimensionless surface tension. Inspecting (2.22) to (2.24) shows that the resulting sets of dimensionless numbers differ, and that different limits can be accessed with the different scalings. In the following we will always use the viscous scaling, i.e. we have as momentum equations

$$\begin{aligned}
u_t + uu_x + wu_z &= -p_x + u_{xx} + u_{zz} + G\sin\alpha & (2.25) \\
w_t + uw_x + ww_z &= -p_z + w_{xx} + w_{zz} - G\cos\alpha. & (2.26)
\end{aligned}$$

For the temperature field we obtain the dimensionless equation

$$\frac{U_0\varrho c_p \Delta T}{l}(T_t + uT_x + wT_z) = \frac{\kappa\Delta T}{l^2}(T_{xx} + T_{zz})$$

and after introducing the Péclet number

$$\mathrm{Pe} = \frac{U_0 l \varrho c_p}{k_{th}}$$

one gets

$$\mathrm{Pe}\,(T_t + uT_x + wT_z) = T_{xx} + T_{zz}. \qquad (2.27)$$

Using again the viscous velocity scale the Péclet number becomes the Prantdl number $\mathrm{Pr} = \eta c_p/k_{th} = \nu/\kappa_{th}$. Besides the governing equations one also has to derive nondimensional boundary conditions from equations (2.6) to (2.3). Many variants exist in the literature. Here we use viscous scales. Assuming further a linear dependence of surface tension on temperature, $\gamma = \gamma_0 - \gamma_T(T - T_\infty)$, where $\gamma_0 = \gamma(T_0)$ we find for the tangential stress condition

$$(u_z + w_x)(1 - h_x^2) + 2(w_z - u_x)h_x = -\mathrm{Ma}\,(T_x + h_x T_z)(1 + h_x^2)^{1/2} \qquad (2.28)$$

where $\mathrm{Ma} = l\varrho\gamma_T\Delta T/\eta^2$ is the Marangoni number [as in Oron et al. (1997), not as in Joo et al. (1991)]. The normal stress condition is

$$p + \frac{2}{1+h_x^2}[-u_x h_x^2 - w_z + h_x(u_z + w_x)] = -S\frac{h_{xx}}{(1+h_x^2)^{3/2}} - \Pi \qquad (2.29)$$

where $S = 1/\mathrm{Ca} = \gamma_0 l\varrho/\eta^2$ is the dimensionless surface tension (Joo et al., 1991) and Ca is the Capillary number (Oron et al., 1997). Note, that the use of γ_0 in the Laplace pressure term is only valid for $\gamma_0 \gg \gamma_T\Delta T$. The free surface boundary condition for the temperature field writes

$$(T_z - h_x T_x)(1 + h_x^2)^{-1/2} + \mathrm{Bi}\,T = 0 \qquad (2.30)$$

where $\mathrm{Bi} = l\alpha_{th}/\kappa$ is the Biot number. The scaling of the remaining boundary conditions is trivial.

2.6 Long-Wave Scaling

The next step is crucial for the derivation of a simplification of the Navier-Stokes equations for the thin film geometry. It makes use of the observation that all length scales parallel to the substrate L, like for instance, periods of surface waves or drop length, are large compared to the film thickness l, i.e.

$$l \ll L \quad \rightarrow \quad \epsilon = l/L$$

where we have introduced the smallness parameter ϵ. This allows to replace the scale l in (2.18) by two different scales for the x and the z coordinate, i.e.

$$x = Lx' = \frac{l}{\epsilon}x', \qquad z = lz'$$

As a consequence of continuity also the velocity components are scaled differently using

$$u = U_0 u', \qquad w = \epsilon U_0 w'$$

In view of the kinematic boundary condition one also has to scale time as

$$t = \frac{L}{U_0} = \frac{l}{\epsilon U_0}t'.$$

Rescaling equations (2.25)-(2.30) yields the transport and continuity equations

$$\begin{aligned}
\epsilon(u_t + uu_x + wu_z) &= -\epsilon p_x + \epsilon^2 u_{xx} + u_{zz} + G\sin\alpha \\
\epsilon^2(w_t + uw_x + ww_z) &= -p_z + \epsilon^3 w_{xx} + \epsilon w_{zz} - G\cos\alpha \\
\epsilon\,\mathrm{Pe}(T_t + uT_x + wT_z) &= \epsilon^2 T_{xx} + T_{zz} \\
u_x + w_z &= 0
\end{aligned} \qquad (2.31)$$

and the boundary conditions at $z = h(x)$

$$(u_z + \epsilon^2 w_x)(1 - \epsilon^2 h_x^2) + 2\epsilon^2(w_z - u_x)h_x = -\mathrm{Ma}\,\epsilon(T_x + h_x T_z)(1 + \epsilon^2 h_x^2)^{1/2}$$

$$p + \frac{2}{1 + \epsilon^2 h_x^2}[-\epsilon^3 u_x h_x^2 - \epsilon w_z + \epsilon h_x(u_z + \epsilon^2 w_x)] = -\frac{1}{\mathrm{Ca}}\frac{\epsilon^2 h_{xx}}{(1 + \epsilon^2 h_x^2)^{3/2}}$$

$$(T_z - \epsilon^2 h_x T_x)(1 + \epsilon^2 h_x^2)^{-1/2} + \mathrm{Bi}\,T = 0$$

$$w = h_t + u h_x$$

$$(2.32)$$

The length scales l and L are still not specified. The mean film thickness and some wave period may be used but other candidates may also be suitable. The detailed scaling depends on the problem studied. However, we must always check that the scales fulfill $l \ll L$.

2.7 Small Inclination or Horizontal Substrate

All fields could now be written as series in ϵ allowing us to solve equations (2.31) to (2.32) order by order. Such a procedure leads at $O(\epsilon)$ to the Benney equation (Benney, 1966; Joo et al., 1991). It has, however, the disadvantage that it explicitly contains the smallness parameter ϵ, i.e. it consists of terms of different order. An alternative approach studies a situation where all physically interesting effects enter the lowest order equations. Mathematically this is achieved by rescaling the fields and dimensionsless numbers as follows (Oron and Rosenau, 1992).

For a small inclination ($\alpha \ll 1$) one introduces a new $O(1)$ variable $\alpha' = \alpha/\epsilon \approx \sin(\alpha)/\epsilon$. This implies $\sin\alpha \to \epsilon\alpha'$ and $\cos\alpha \to 1 - O(\epsilon^2)$. Furthermore one chooses $\mathrm{Ca}' = \mathrm{Ca}/\epsilon^2$ and accounts for the fact that for small inclinations all velocities are small, i.e. a new $\vec{v'} = \vec{v}/\epsilon$ is introduced. After dropping the dashes this leads in lowest order in ϵ to the transport equations

$$u_{zz} = p_x - G\alpha, \tag{2.33}$$
$$p_z = -G, \tag{2.34}$$
$$T_{zz} = 0; \tag{2.35}$$

the continuity equation $u_x + w_z = 0$ and the boundary conditions at $z = 0$

$$u = w = 0, \tag{2.36}$$
$$T = 1; \tag{2.37}$$

and at $z = h(x)$

$$w = \partial_t h + u\partial_x h, \tag{2.38}$$
$$u_z = -\mathrm{Ma}\,(T_x + h_x T_z), \tag{2.39}$$
$$p = -\frac{h_{xx}}{\mathrm{Ca}} - \Pi(h), \tag{2.40}$$
$$T_z = -\mathrm{Bi}\,T; \tag{2.41}$$

respectively. The resulting system can be readily solved. One obtains the fields

$$p(x,z) = G(h-z) - \Pi(h) - \frac{h_{xx}}{\mathrm{Ca}}, \qquad (2.42)$$

$$T(x,z) = 1 - \frac{\mathrm{Bi}}{(1+\mathrm{Bi}\,h)}z, \qquad (2.43)$$

$$\text{and} \quad u(x,z) = \left(\frac{z^2}{2} - zh\right)(p_x - G\alpha) + \frac{\mathrm{MaBi}\,h_x}{(1+\mathrm{Bi}\,h)^2}z. \qquad (2.44)$$

The kinematic boundary condition and continuity give

$$\partial_t h = -\partial_x \Gamma \qquad (2.45)$$

where $\Gamma = \int_0^h u\,dz$ is the flow in the laboratory frame.

2.8 The Film Thickness Evolution Equation

Finally, using equations (2.44) and (2.42) in equation (2.45) gives the evolution equation for the thickness profile of a film on a slightly inclined substrate

$$\partial_t h = -\partial_x \left\{ \frac{h^3}{3}\left[\partial_x\left(\frac{h_{xx}}{\mathrm{Ca}} - Gh - \Pi(h)\right) + \frac{3}{2h}\frac{\mathrm{Ma\,Bi}\,h_x}{(1+\mathrm{Bi}h)^2} + G\alpha\right]\right\}. \qquad (2.46)$$

By a further rescaling we eliminate Ca and write in a more general form

$$\partial_t h = -\partial_x \left\{Q(h)\,\partial_x\left[\partial_{xx}h - \partial_h f(h)\right] + \chi(h)\right\} \qquad (2.47)$$

where $Q(h)$ is a mobility factor and $\chi(h)$ stands for a generalized driving parallel to the substrate. The use of slip boundary conditions at the substrate leads to different Q and χ but has no effect on the equation otherwise (Oron et al., 1997). The term in square brackets represents the negative of a pressure consisting of the Laplace term $\partial_{xx}h$ and an additional contribution $-\partial_h f(h)$ written as the derivative of a local free energy. The latter has to be specified for the respective studied problem. For the system presented here we have

$$\partial_h f(h) = -\Pi - Gh - \frac{3}{2}\mathrm{Ma\,Bi}\int \frac{dh}{h(1+\mathrm{Bi}\,h)^2}.$$

For inhomogeneous systems, $f(h)$ may depend also on position. Equation (2.47) can be written in a variational form

$$\partial_t h = \partial_x\left\{Q(h)\,\partial_x\frac{\delta F[h]}{\delta h}\right\} - \partial_x \chi(h) \qquad (2.48)$$

with $\delta/\delta h$ denoting functional variation with respect to h and the free energy functional

$$F[h] = \int\left[\frac{1}{2}(\partial_x h)^2 + f(h)\right]dx. \qquad (2.49)$$

Without lateral driving, i.e. for $\chi = 0$, equation (2.48) corresponds to the simplest possible equation for the dynamics of a conserved order parameter field (Langer, 1992). In this case the system is variational (or relaxational) and $F(h)$ can be used as a Lyapunov functional because it fulfills $dF/dt \leq 0$ (Langer, 1992; Mitlin, 1993; Oron and Rosenau, 1992). A prominent representative of this class of systems is the Cahn-Hilliard equation describing the evolution of a concentration field for a binary mixture (Cahn, 1965).

For $\chi \neq 0$, equation (2.48) does not represent a variational system any more, the dynamics is non-relaxational, i.e. $F[h]$ does not necessarily decrease in time. Typical examples are the driven (or convective) Cahn-Hilliard equation (Golovin et al., 2001), the Kuramoto-Sivashinsky equation (Kevrekidis et al., 1990; Kuramoto and Tsuzuki, 1976; Sivashinsky, 1977) and the Benney equation and its variants (Benney, 1966; Gjevik, 1970; Joo et al., 1991; Lin, 1974; Pumir et al., 1983; Scheid et al., 2005).

2.9 The Additional Pressure Term

Equations like (2.47) can be used to study a wide variety of systems (Oron et al., 1997). Here we focus on two physical situations: ultrathin films below $100\,\mathrm{nm}$ thickness on partially wettable substrates that undergo dewetting (Bischof et al., 1996; Brochard-Wyart and Daillant, 1989; de Gennes, 1985; Reiter, 1992; Thiele et al., 1998) and thin films below about $100\,\mu m$ thickness heated from below. They may be unstable due to a long-wave Marangoni instability (Burelbach et al., 1988; Oron, 2000; Oron and Rosenau, 1992; VanHook et al., 1995, 1997). In the former case the additional pressure $-\partial_h f(h)$ in equation (2.47) corresponds to the disjoining pressure $\Pi(h)$ introduced above in Section 2.4 [$f(h)$ corresponds to $V(h)$]. Two forms for the pressure used here are

$$\partial_h f(h) \;=\; 2\kappa\,e^{-h}\left(1 - e^{-h}\right) + Gh \tag{2.50}$$

$$\text{and} \quad \partial_h f(h) \;=\; \pm(-\frac{b}{h^3} + e^{-h}). \tag{2.51}$$

The first term of the form (2.50) was derived by Pismen and Pomeau (2000) combining the long wave approximation for thin films (Oron et al., 1997) with a diffuse interface model for the liquid-gas interface (Anderson et al., 1998). Also density variations close to the solid substrate enter their calculation in a natural way. The second term accounts for the hydrostatic pressure. Equation (2.50) has the advantage of remaining finite even for vanishing film thickness. The parameter G is normally very small, however, it is possible to study the qualitative system behavior using a small but not very small G.

Equation (2.51) combines a long-range apolar van der Waals interaction and a short-range polar (electrostatic or entropic) interaction (Sharma, 1993a,b) as discussed above in Section 2.4. If the positive sign is used, both disjoining pressures lead to qualitatively very similar results (Thiele et al., 2002b).

There exist various other forms of disjoining pressures used in the literature. Most results are similar if the pressure in question describes a partially wetting situation and allows for a stable precursor film. Nondimensionalized examples for combinations of power laws are, for instance, $\partial_h f(h) = 1/h^3 - b_1/h^6$ derived from a diffuse interface theory with nonlocal interactions (Pismen, 2001; Pismen and Pomeau, 2004; Pismen and Thiele, 2006) and $\partial_h f(h) = 1/h^3 - b_1/(h + d_{co})^3 - b_2/h^9$ describing a film on a substrate with a coating of thickness d_{co} (Sharma and Reiter, 1996; Seemann et al., 2005). The parameters b_i are dimensionless constants.

For the heated film the additional pressure $\partial_h f(h)$ results from the tangential stress condition at the free surface (Oron and Rosenau, 1992). It is given by

$$\partial_h f(h) = Gh - \frac{3}{2}\,\mathrm{Bi}\,\mathrm{Ma}\left[\log\left(\frac{h}{1+\mathrm{Bi}\,h}\right) + \frac{1}{1+\mathrm{Bi}\,h}\right]. \tag{2.52}$$

Besides these basic situations, combinations of the different pressure terms can also be used. For example, supplementing equation (2.52) with a stabilizing apolar interaction allows to study the long-time coarsening of a heated thin film (Bestehorn et al., 2003). To account for a chemically inhomogeneous substrate one can use a disjoining pressure that varies in space (Konnur et al., 2000; Thiele et al., 2003). For an inhomogeneous heating one modulates the Marangoni number in equation (2.52) (Scheid et al., 2002; Skotheim et al., 2003).

2.10 Analysis Techniques

The analysis of the thin film equation (2.47) in the four different geometries consists of: (1) analytical determination of fixed points, i.e. flat film solutions, and of their stability properties; (2) calculation of stationary solutions[6], their bifurcations and linear stability using continuation techniques; and (3) numerical integration in time. The majority of the presented results is based on continuation techniques (Doedel et al., 1991a,b, 1997).

With continuation techniques one can obtain solutions of a problem for a certain set of control parameters by iterative techniques from known solutions nearby in the parameter space. Specifically, the film thickness equation is written as a dynamical system, i.e. as a system of ordinary differential equations

$$u'(x) = f(u(x), p) \qquad \text{with} \qquad f, u\,in\,\mathbf{R}^n \tag{2.53}$$

where the dash indicates derivation with respect to x and p denotes the set of control parameters. Here we present the basic steps of numerical continuation of the solution of thin film equations by using the package AUTO (Doedel et al., 1997). Equation (2.53) subject to boundary and integral conditions is discretized in space. The resulting algebraic system is solved iteratively. AUTO uses the method of Orthogonal Collocation for discretizing solutions, where the solution is approximated by piecewise polynomials with 2–7 collocation points per mesh interval. The mesh is adaptive as to equidistribute the discretization error. Starting from known solutions AUTO then tries to find nearby solutions to the discretized system, by using a combination of Newton and Chord iterative methods. Once the solution has converged AUTO proceeds along the solution branch by a small step in the parameter space defined by the free continuation parameters and restarts the iteration.

Boundary conditions and/or integral conditions require additional free parameters which are determined simultaneously and are part of the solution to the differential equation. The package AUTO is limited to the continuation of ordinary differential equations (ODE's), thus it can only be used to compute droplet solutions in two dimensions. As an example we consider the continuation of stationary surface waves or sliding drops steady

---

[6]Stationary solutions are solutions that are steady in some co-moving frame of reference.

in a comoving frame. After transforming equation (2.47) into the comoving frame with velocity v and integrating the resulting time-independent thin-film equation we have the system of ODE's

$$\begin{aligned}
u_1' &= u_2 \\
u_2' &= u_3 \\
u_3' &= -\frac{\chi(u_1) + v u_1 - C_0}{Q(u_1)} + \partial_{u_1} f(u_1)
\end{aligned} \qquad (2.54)$$

where C_0 is an integration constant (see Section 5). u_1, u_2 and u_3 denote h, $\partial_x h$ and $\partial_{xx} h$, respectively. The volume is conserved, thus we need to specify the integral condition

$$0 = \frac{1}{L} \int u_1 \, dx - \bar{h} \qquad (2.55)$$

where L is the system length and $\bar{h}$ is the mean film thickness. We also use periodic boundary conditions

$$\begin{aligned}
u_1(0) &= u_1(L) & (2.56) \\
u_2(0) &= u_2(L) & (2.57) \\
u_3(0) &= u_3(L). & (2.58)
\end{aligned}$$

Translation of a solution along the substrate yields another valid solution. However, continuation along the 'trivial' families of solutions related through continuous symmetries of the system is not wanted. It is effectively forbidden by a pinning condition in the form of an additional boundary or integral condition. The number of boundary (NBC) and integral conditions (NINT) determines the number of parameters (NPAR), that are varied during the continuation process. Specifically NPAR=NBC+NINT-NDIM+1, where NDIM is the dimensionality of (2.53), i.e. here NDIM= 3. The resulting three free parameters are called continuation parameters and may, for instance, be the system size L, the integration constant C_0 and the velocity v.

The challenge is often to provide a starting solution for the continuation. For surface waves or sliding drops one can use analytically known surface waves obtained by a linear perturbation of the flat film. In other cases the initial solution is given numerically. Finally note, that continuation is not only used to determine families of stationary solutions (Thiele et al., 2001c,b), but also to follow their stability behavior (Thiele and Knobloch, 2003) and bifurcations, like saddle-node bifurcations or branching points (Thiele and Knobloch, 2004; Scheid et al., 2005). This can be used to determine phase diagrams mapping the existence and stability of various solution types (John et al., 2005).

In general, the details of the adopted techniques are different for different geometries. These are introduced were needed in the course of the following sections that analyze specific physical situations corresponding to the different geometries shown in Figure 1. However, many practically relevant thin film systems can not be modeled by a single evolution equation of the type of equation (2.47). Therefore, in Section 7 we extend the theory to situations where two order parameter fields have to be taken into account.

3 Horizontal Homogeneous Substrate

3.1 General Analysis

For the horizontal homogeneous substrate there exists no lateral driving force, i.e.
$\chi(h) = 0$. Equation (2.47) becomes

$$\partial_t h = -\partial_x \left\{ Q(h)\, \partial_x \left[\partial_{xx} h - \partial_h f(h) \right] \right\}. \tag{3.1}$$

Steady solutions. To study steady solutions we set $\partial_t h = 0$ and integrate equation (3.1) to get

$$0 = \partial_{xxx} h - \partial_{hh} f(h)\, \partial_x h - \frac{C_0}{Q(h)}. \tag{3.2}$$

We look for periodic arrays of drops or holes, localized drops or holes, and flat film solutions. The reflection symmetry $(x \to -x)$ with respect to the extrema of the solutions or the flat film at infinity, respectively, implies $C_0 = 0$. Another way to see this is to interpret C_0 as the mean flow. Without an additional driving force no mean flow occurs, i.e. $C_0 = 0$. A second integration yields

$$0 = \partial_{xx} h(x) - \partial_h f(h) + C_1. \tag{3.3}$$

The constant C_1 accounts for external conditions like chemical potential, vapor pressure or mass conservation. For the latter case that we will focus on here, C_1 takes the role of a Lagrange multiplier. The choice $C_1 = \partial_h f(h_0)$ ensures the flat film $h(x) = h_0$ to be a solution. Note that equation (3.3) also follows directly from the minimization of the energy functional (2.49) and mass conservation. Later we will compute periodic solutions of equation (3.3) and parameterize them by their mean film thickness, $\bar{h} = (1/L) \int_0^L h(x)dx$, and period, L.

In a two-dimensional situation, equation (3.3) is equivalent to the equation of motion for a particle in a potential $V(h) = -f(h) + C_1 h$. For radial-symmetric solutions in a three-dimensional situation it is equivalent to an equation of motion including a time-dependent friction term (see Deissler and Oron (1992); Thiele et al. (2002b); Bestehorn et al. (2003)).

Besides the chosen h_0, other flat film solutions or fixed points of equation (3.3), $h(x) = h_f$, may exist. They are given by

$$\partial_h f(h_f) = \partial_h f(h_0). \tag{3.4}$$

For the disjoining pressure shown in Figure 5 [Equation (2.17)] one finds one or three fixed points depending on the value of b and h_0. The bifurcation points between the two regimes are given by $\partial_{hh} f(h_0) = 0$ and equation (3.4). Linearizing equation (3.3) at the fixed points one finds that for $\partial_{hh} f(h_f) > 0$ (< 0) they are saddles (centers). This corresponds to the results of the linear stability analysis for flat films shown next.

In order to determine thickness profiles that are solutions of equation (3.3) we multiply (3.3) by h_x and integrate once to obtain

$$\partial_x h = \sqrt{2}\, \sqrt{f(h) - C_1 h - C_2}. \tag{3.5}$$

where C_2 is a constant discussed below.

Linear stability of flat films. To assess the linear stability of the flat film $h(x) = h_0$ we use a Fourier mode decomposition $h(x) = h_0 + \epsilon \exp(\beta t + ikx)$ where β and k are the growth rate and the wave number of the disturbances, respectively. Linearizing the full time-dependent equation (3.1) for $\epsilon \ll 1$ yields the dispersion relation

$$\beta = -Q(h_0)\, k^2 \left[k^2 + \partial_{hh} f(h_0) \right]. \tag{3.6}$$

The possible outcomes are sketched in Figure 7. The flat film is unstable for $\beta > 0$, i.e.

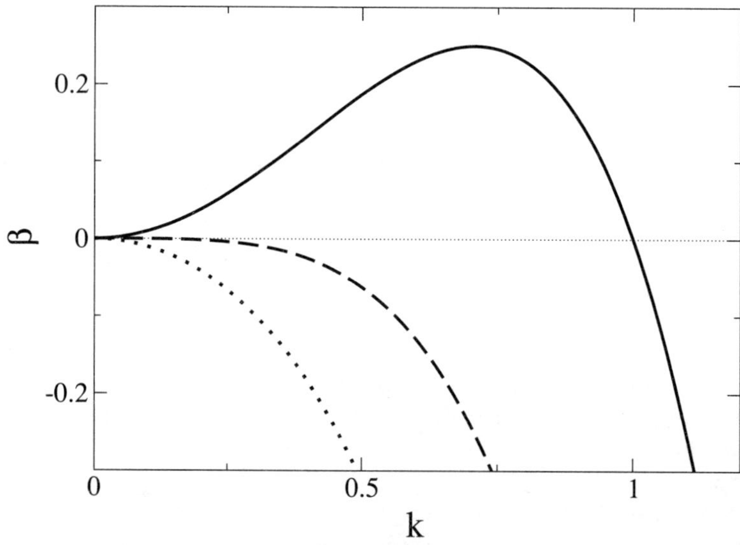

Figure 7. Dispersion relations for a flat film on a horizontal homogeneous substrate. They correspond to the unstable (solid line), marginally stable (dashed line) and linearly stable case (dotted line).

there exists some range of unstable wavenumbers $0 < k < k_c$ for

$$\partial_{hh} f(h_0) < 0. \tag{3.7}$$

The instability has its onset at $\partial_{hh} f = 0$ with $k_c = 0$. This is also called the 'marginally or neutrally stable case'. Because of the onset at zero wavenumber it is called a long-wave instability. It corresponds to an instability of type II_s in the classification of Cross and Hohenberg (1993) (see their section IV.A.1.b) The critical wave number is

$$k_c = \sqrt{-\partial_{hh} f(h_0)} \tag{3.8}$$

and the corresponding wavelength is $\lambda_c = 2\pi/k_c$. The thickness profile $h(x) = h_0 + \epsilon \exp(ik_c x)$ is neutrally (or marginally) stable ($\beta = 0$) and represents a small amplitude periodic steady solution of equation (3.1).

For a linearly unstable film the fastest growing mode has the wavelength $\lambda_m = \sqrt{2}\lambda_c$ and the growth rate

$$\beta_m = \frac{1}{4}\, Q(h_0)\, [\partial_{hh} f(h_0)]^2. \tag{3.9}$$

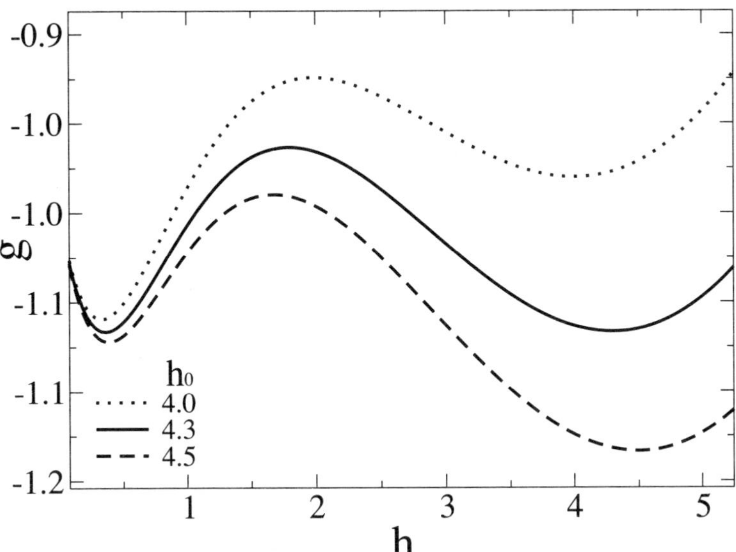

Figure 8. The local energy $g(h) = f(h) - C_1(h_0)\,h$. The disjoining pressure is given by (2.50) (see Thiele et al. (2001c)). An equivalent plot for the disjoining pressure (2.51) can be found in Thiele et al. (2001a).

Nonlinear stability. A linearly stable flat film may not be absolutely stable. It can be unstable to finite amplitude disturbances corresponding to a metastable flat film. This may indicate the occurrence of a sub-critical instability. Only if for a given mean film thickness there is no thickness profile with smaller energy, the flat film is absolutely stable. To further clarify this issue we assume an infinitely long film of thickness h_0. Only a small part of finite length s has a different thickness h to ensure that the mean film thickness remains h_0. The width of the finite transition region between the two thicknesses is small compared to s, so its energy can be neglected. Now we can calculate the energy per unit length of the changed part

$$g(h) = f(h) - C_1(h_0)\,h. \tag{3.10}$$

The function $g(h)$ is plotted in Figure 8 for different values of h_0. The two minima represent a lower and an upper linearly stable film thickness. However, only the deeper minimum corresponds to an absolutely stable state, whereas the other one is metastable. The maximum represents a linearly unstable film thickness. There exist an upper and a lower limit of the metastable thickness range, denoted by h_m^u and h_m^d, respectively. They are characterized by minima of equal depth in Figure 8 and identical values of $C_1(h_m)$, i.e. by

$$\begin{aligned}
\partial_h f|_{h_m^u} &= \partial_h f|_{h_m^d} \\
g(h_m^u) &= g(h_m^d).
\end{aligned} \tag{3.11}$$

Note, that equations (3.11) are equivalent to a Maxwell construction (see also Mitlin (1993); Samid-Merzel et al. (1998); Mitlin (2000)).

Using the results for the onset of the linear instability of the flat film equation (3.7) and equations (3.11) for its absolute stability one can calculate the stability diagram. An example is given below in Figure 12. These diagrams are valid for two- and three-dimensional film geometries.

Equilibrium contact angle. If equation (3.11) yields two equilibrium thicknesses, i.e. binodal thicknesses, one can not define a mesoscopic equilibrium contact angle in an asymptotic sense (Pismen and Thiele, 2006) because no solution profile continues towards an infinite height. However one can still define equilibrium contact angles as the slope at the inflection point of the profile. This situation is always encountered for the disjoining pressure derived from diffuse interface theory when including gravity [Equation (2.50)] and for $b > 8/e^2$ also for the apolar/polar combination equation (2.51).

A simpler result is obtained if only one finite equilibrium film thickness exists (the second one is infinite) as for the disjoining pressure (2.51) for $b < 8/e^2$ and always for (2.50) without gravity. Then a droplet of infinite height coexists with a precursor film of thickness h_p given by $\partial_h f(h_p) = 0$. One also finds $C_1 = 0$ and $C_2 = f(h_p)$ [see equations (3.3) and (3.5)]. Going back to physical (dimensional) variables the asymptotic equilibrium contact angle given as h_x for $h \to \infty$, is $\theta_e = \sqrt{-2f(h_p)/\gamma}$.

Periodic solutions. In order to study the various non-constant thickness profiles, i.e. periodic (assemblies of drops or holes), homoclinic (single drops or holes) and heteroclinic (fronts) solutions of equation (3.5) we choose for the integration constant

$$C_2 = f(h_m) - (\partial_h f(h_0)) h_m \tag{3.12}$$

where h_m is the maximal or minimal thickness for periodic solutions. For localized solutions $h_m = h_0 = h_\infty$. Hence, all solutions are parameterized by the pair (h_0, h_m) or (C_1, C_2). Equation (3.5) allows to plot the solutions in the phase plane (h, h_x) [see Mitlin (1993); Thiele et al. (2001c)]. Three qualitatively different phase portraits (each for fixed C_1) can be observed (see Figure 9). They represent drop, hole and front regimes, respectively. In the hole [drop] regime one finds beside periodic solutions a homoclinic solution representing a localized hole [drop] in an infinitely extended flat film (shown on the lower line of Figure 9). These localized profiles can be found in the metastable range for flat films.

In the front regime, besides the periodic solutions, one also finds two heteroclinic solutions that connect the lowest and the highest fixed point, thus representing localized front or kink solutions that connect two infinitely extended flat films of thicknesses, h_m^d and h_m^u (right image on lower line of Figure 9). The fronts exist only on the border between metastable and stable flat films.

In the following we concentrate on the periodic solutions. They exist for parameter ranges corresponding to linearly unstable or metastable flat films. Depending on the relevant control parameters, like mean film thickness, Marangoni number and interaction constants, one can distinguish three qualitatively different families of solutions. Figure 10

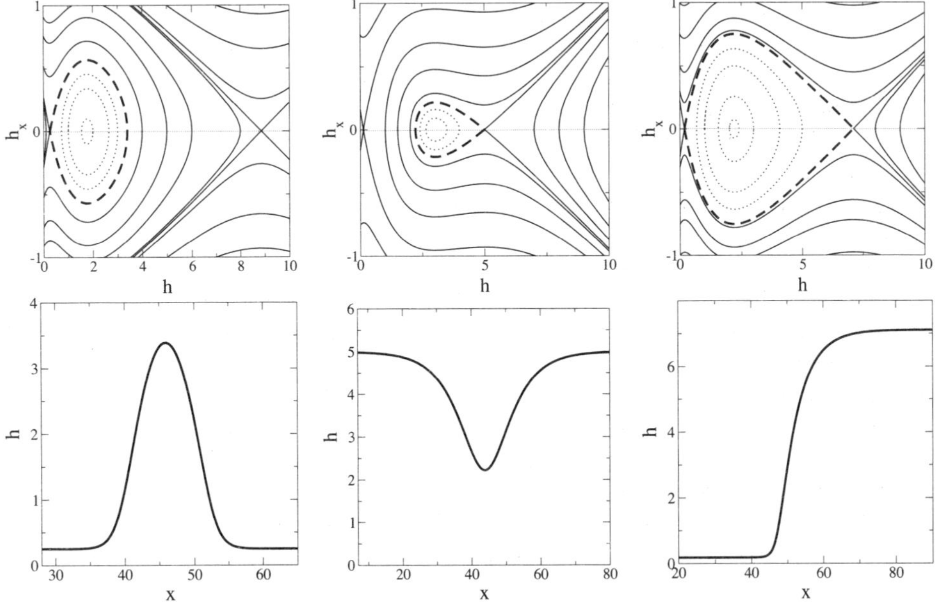

Figure 9. Sketches of qualitatively different phase portraits in the $(h, \partial_x h)$ plane (upper line) and corresponding localized profiles (lower) for drop (left), hole (middle) and front regime (right). For details on the parameters see Thiele et al. (2001c).

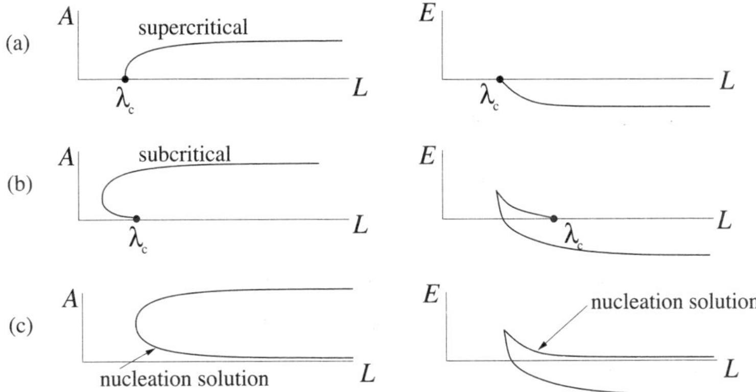

Figure 10. Sketches of the three types of families of periodic solutions. In (a) and (b) a branch bifurcates supercritically and subcritically, respectively, from the unstable flat film solution. In (c) two branches exist that both continue towards infinite period. The flat film is metastable. In (b) and (c) the low-energy branch corresponds to nucleation solutions.

sketches the different types of solutions when the period is chosen as the main control parameter and the mean film thickness is fixed. The solution families are characterized by the amplitude (left) and energy (right) of the solutions. The energy E corresponds to the functional F (Equation (2.49)) per length normalized by the energy of the flat film.

(a) Deep inside the linear unstable film thickness range there exists only one branch of periodic solutions. It bifurcates supercritically from the flat film solution at $L = \lambda_c = 2\pi/k_c$ (Equation (3.8)) and continues towards infinite period. The amplitude/energy increases/decreases monotonically with increasing period. The energy is always lower than that of the corresponding flat film.

(b) Closer to the metastable range but still for linearly unstable film thicknesses there exist two branches of solutions. For both, the energy decreases with increasing period. The high energy branch bifurcates at $L = \lambda_c$ subcritically from the flat film solution. Its energy is always higher than the one of the flat film. This branch is linearly unstable and represents nucleation solutions that have to be 'overcome' if the film is to break into finite portions with size $p < \lambda_c$. The energy of the lower branch decreases rapidly with increasing period. However, there is a very small range of periods where the flat film has the smallest energy implying its absolute stability for systems of this size. Consequently, for this range the low-energy periodic solution is only metastable.

(c) For metastable flat films there exist two solution branches that both continue towards infinite period. The upper one consists of nucleation solutions of different periods that separate energetically the lower periodic solution from the linearly stable flat film solution.

Linear stability of steady solutions. The linear stability of the periodic solutions is assessed by linearizing the full time dependent equation (3.1) around the periodic solutions, $h_0(x)$. The ansatz

$$h(x,t) = h_0(x) + \epsilon\, h_1(x)\, \exp(\beta t) \tag{3.13}$$

gives a linear eigenvalue problem for the growth rate β and disturbance h_1.

$$\beta h_1(x) = \mathbf{N_0}[h_0(x)]\, h_1(x) \tag{3.14}$$

with

$$\begin{aligned}
\mathbf{N_0} h_1 &= -\partial_x \left\{ Q_h h_1\, \partial_x(\partial_{xx} h_0 - \partial_h f) \right\} \\
&\quad -\partial_x \left\{ Q\, \partial_x(\partial_{xx} h_1 - \partial_{hh} f\, h_1) \right\}.
\end{aligned} \tag{3.15}$$

Q and f and their derivatives are functions of $h_0(x)$. To determine $h_1(x)$ and β by continuation one has to follow simultaneously the steady solution $h_0(x)$ of equation (3.3) and the solution of equation (3.14). Because of the variational structure of the problem all β is real. Therefore β can be directly determined as branching points (Doedel et al. (1997)). The required extended system consists of 7 first order differential equations (3 for h_0 and 4 for h_1).

Next we present some selected results for the dewetting of ultrathin films on homogeneous substrates (Section 3.2) and for the long-wave Marangoni-Bénard instability of heated thin films (Section 3.3).

3.2   Ultrathin Partially Wetting Films

While hydrodynamical surface instabilities in thin-film flows are investigated since the experiments of the Kapitzas (Kapitza, 1949; Kapitza and Kapitza, 1949), in soft matter physics they became increasingly important for the understanding of structure formation in thin films on solid substrates, since the work on dewetting by Reiter only one decade ago (Reiter, 1992). In this paradigmatic experiment a polymer film on a solid substrate is brought above its glass transition temperature, ruptures, and the formed holes grow resulting in a network of liquid rims. The latter may decay subsequently into small drops[7].

The ongoing pattern formation can be described by equation (3.1) using a disjoining pressure $\partial_h f(h)$ appropriate for the combination of materials used and accounting also for eventual coatings. A variety of different combinations of stabilising or destabilising exponentials and power laws are used (Hunter, 1992; Israelachvili, 1992; Teletzke et al., 1988) and still new candidates for underlying physical effects besides dispersion or electrostatic forces are discussed (Pismen and Pomeau, 2000; Schäffer et al., 2003; Schäffer and Steiner, 2002; Ziherl et al., 2000).

One of the controversial questions (Bischof et al., 1996; Brochard-Wyart and Daillant, 1989; Jacobs et al., 1998; Reiter, 1992; Thiele et al., 1998, 2001a; Xie et al., 1998) concerns the mechanism of the rupture of the initially flat film. Does it occur via *instability* or *nucleation*? Most literature studies relate the occurrence of surface instability and heterogeneous nucleation at defects to linearly unstable and metastable films, respectively (see Figure 12 below and the discussion above in Section 3.1).

The importance of this question in dewetting is due to the fact that in most systems the evolution is frozen before significant coarsening occurs, i.e. the mechanism of the initial rupture still determines the structure. On the contrary, for the decomposition of a binary mixture (as described by the Cahn-Hilliard equation) there is little discussion about this point, what seems to be the main interest there is the scaling behavior of the long-time coarsening, because it gives the evolution of the length scales that can be measured experimentally. For thin films coarsening is up to now only of minor interest (exceptions are Mitlin (2001); Bestehorn et al. (2003); Glaser and Witelski (2003); Pismen and Pomeau (2004), see also Section 3.4) because for the experimental systems used in dewetting the time scale for large-scale coarsening is very large.

Recent work has re-evaluated for two-dimensional model systems the two rupture mechanisms starting from an analysis of the solution structure of equation (3.1). As elucidated above at Figure 10, nucleation solutions exist not only in the metastable thickness range but also in a part of the linearly unstable thickness range. These unstable solutions 'organize' the evolution of the thin film by offering a fast track to film rupture that does not exist in their absence.[8]

---

[7]Note, however, that descriptions of the patterns formed in this kind of process can already be found for liquid-liquid dewetting in papers by Tomlinson (Tomlinson, 1870, footnote 18 on p. 40) for turpentine on water and Marangoni (Marangoni, 1871, p. 352f) for oil on water.

[8]For the decomposition of a binary mixture Novick-Cohen (1985) discussed such solutions as an evidence for a smooth transition from spinodal decomposition to nucleation somewhere within the classical spinodal.

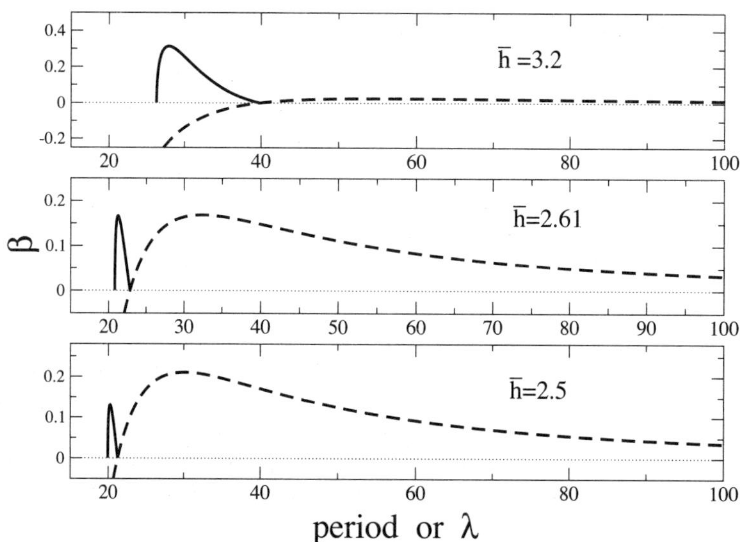

Figure 11. Comparison of the linear growth rates of the surface instability of the flat film (dashed lines) and of the linear unstable nucleation solutions (solid lines). In the first (last) image the nucleation solutions (flat film modes) dominate whereas in the intermediate image both have equal maximal growth rate (see Thiele et al. (2001c)).

For a linearly unstable film with a solution structure as in Figure 10 (b) the nucleation solutions are unstable solutions that have to be overcome to break a film in smaller portions than the critical linear wavelength, λ_c. However, they also influence the local rupture dynamics if there exist localized disturbances of the film surface (defects) with lateral extensions smaller than λ_c. Then, locally the nucleation solutions first attract the evolution to later repel it with a well-defined rate β_{nuc}. The rate can be obtained by analyzing the linear stability of the nucleation solutions. This property reflects the fact that the nucleation solutions are saddles in the space of all possible surface profiles.

To evaluate the influence of this nucleation solution mediated rupture one has to compare the related rates β_{nuc} and the rates of the 'normal' linear rupture modes of a flat unstable film given by equation (3.6). Such comparisons are shown in Figure 11 for different mean film thicknesses. The heights of the two maxima now allow to predict whether defects have an influence on the resulting morphology or not.

As a result one distinguishes 'nucleation-dominated' and 'instability-dominated' behavior for linearly unstable films as indicated in Figure 12 (Thiele et al., 2001a). The new boundary separating the two sub-ranges is defined as the line where the two maxima in Figure 11 have the same height.

If the behavior is nucleation-dominated an initial disturbance grows much faster than the active linear instability of the flat film (Figure 13 (a)). The produced holes expand and if the dynamically produced surface depression just outside the rim becomes larger than the respective nucleation solution, secondary nucleation events occur. The resulting

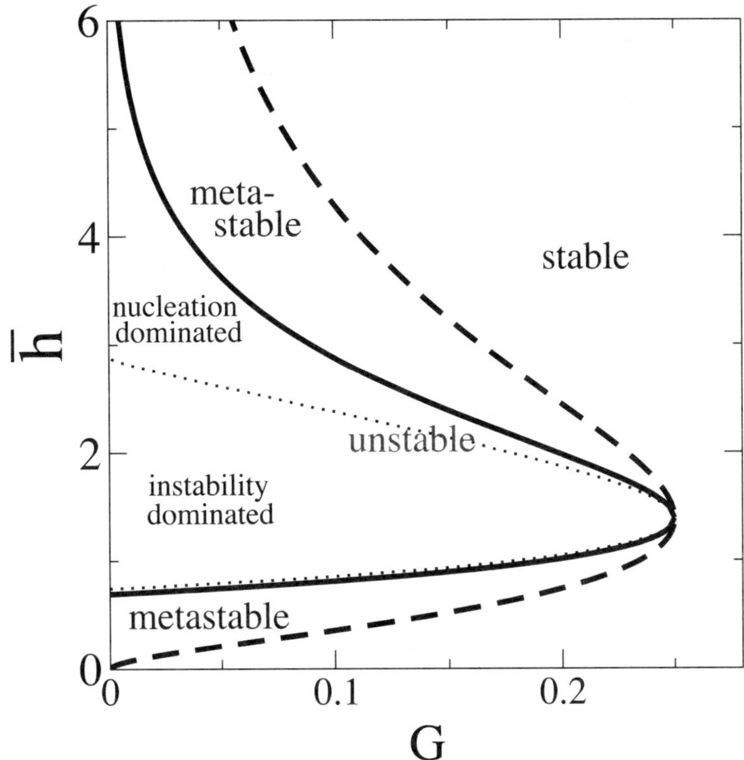

Figure 12. Stability diagram for a thin liquid film including the binodal (dashed line) and the spinodal (solid line) for the disjoining pressure of equation (2.50). The dot-dashed line indicates the boundary between nucleation-dominated and instability-dominated sub-ranges within the linearly unstable range.

structure is a set of holes with distances unrelated to λ_m. It depends strongly on the properties of the initial defect. If the behavior is instability-dominated the initial disturbance also starts to grow and acts as starting point for the most unstable flat film mode (Figure 13 (b)). Undulations of period λ_m extend laterally to give finally a periodic set of holes nearly independent of the initial perturbation.

This qualitative result neither depends on the details of the used disjoining pressure (Thiele et al., 2002b) nor is it expected to be different in three-dimensional systems. Therefore our results (Thiele et al., 2001a) and Thiele et al. (2001c) explain qualitatively *why* the morphological transition shown by Becker et al. (2003) occurs. For a detailed analysis of the result of Becker et al. (2003) and application of our findings to further experiments (Du et al., 2002; Meredith et al., 2000) see Thiele (2003a,b)[9].

---

[9]Note that in Seemann et al. (2005) our result was misunderstood (p. S285/S286). The predicted boundary between the two qualitatively different sub-ranges lies well inside the range

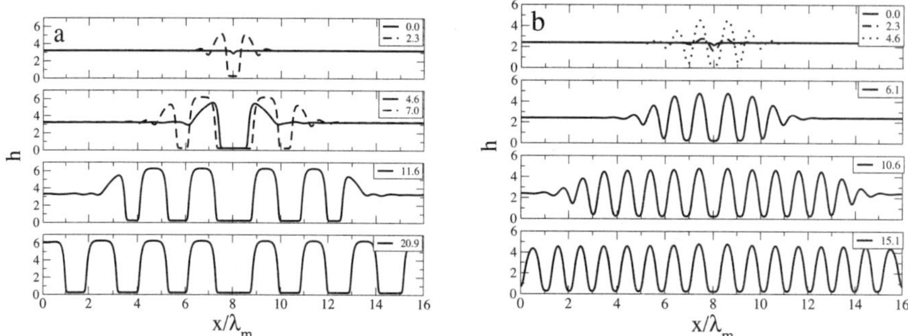

Figure 13. Snapshots of the time evolution in the nucleation-dominated (a) and instability-dominated (b) sub-ranges within the linearly unstable film thickness range. The insets give the evolution time and x is measured in units of the fastest growing wavelength (for details see Thiele et al. (2001c)).

3.3 Heated Thin Films

The second case where the theoretical framework presented in Section 3.1 can be applied are large scale surface deformations of thin liquid films on a smooth solid homogeneously heated substrate. Without heating the film is stable. However, this is no longer the case once thermocapillary (Marangoni) effects are included. The resulting instability evolves according to equation (3.1) with the additional pressure given by equation (2.52) (Burelbach et al., 1988; Oron and Rosenau, 1992; Thiele and Knobloch, 2004). We consider only parameter regimes where the short-wave convective mode (Davis, 1987; Golovin et al., 1994) does not occur. The details of the thermocapillary instabilities depend on the assumed dependence of the surface tension on temperature (Nepomnyashchy et al., 2002). Usually it is taken to be linear (Deissler and Oron, 1992). A similar equation for a film below an air layer of finite thickness, was given by VanHook et al. (1997) in connection with their investigation of the formation of dry spots. The study by Boos and Thess followed numerically the evolution of a film profile towards rupture using the full Stokes equation in combination with a linear temperature field (Boos and Thess, 1999), and identified a cascade of consecutive 'structuring events' pointing towards the formation of a set of drops as the final state of the system.

Thiele and Knobloch (2004) revisited the problem of a heated thin film on a horizontal and a slightly inclined substrate to study the multiplicity of solutions to the nonlinear evolution equation and their stability properties. They also investigated the effects of a small inclination of the substrate (see Section 5). The basic behavior is captured by a simplified model that omits complications due to effective molecular interactions, i.e. a model without disjoining pressure. One finds that the possible two-dimensional steady solutions of equation (3.1) can be determined independently of the fact that they cannot

---

of linearly unstable film thicknesses, i.e. the film is not metastable as stated there. Our theory even explains why 'this behaviour should be typical for thin films of Newtonian liquids particularly in the unstable state' (Seemann et al., 2005).

be reached from the initial condition of a flat film by integration in time. As for the ultrathin films studied above in Section 3.2, one encounters unstable nucleation solutions and drop-like solutions. A difference is that the primary instability is always subcritical, i.e. defects always play an important role.

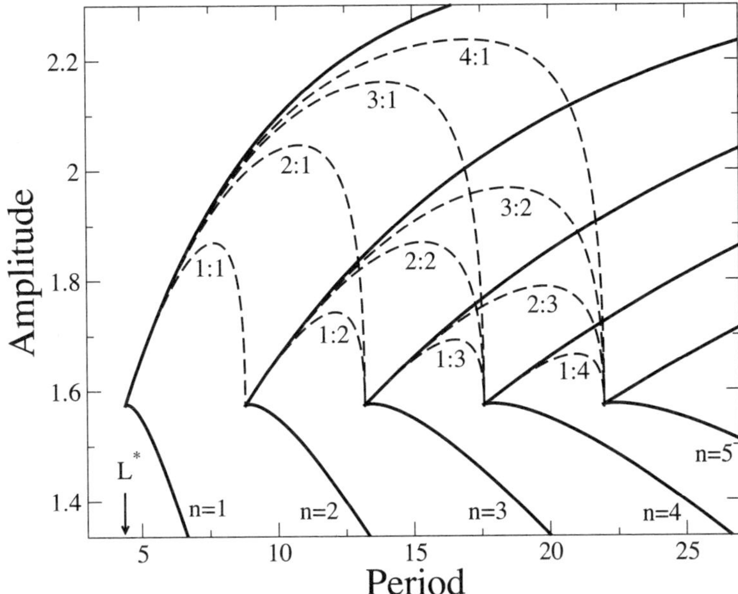

Figure 14. The amplitude $\Delta h \equiv h_{max} - h_{min}$ along single drop and multidrop branches with microscopic contact angle $\theta_0 = 0$ as a function of period L. They are steady solutions of equation (3.3) with (2.52) with Ma = 3.5, Bi = 0.5 and Bo = 1.0. The solid lines show the primary $n = 1, \ldots, 5$ branches where n gives the number of drops per period. They consist of periodic nucleation solutions (lower part) and the periodic drop solutions (upper part). The dashed lines show the different possible multidrop branches with maximal internal symmetry (i.e. branch $i : j$ has i identical drops of one type and j identical drops of another type) and no dry holes. Multidrop solutions with broken internal symmetry are present between these multidrop branches as described by Thiele and Knobloch (2004). Every multidrop solution of this type in turn represents the starting point for $i + j$ branches containing finite dry holes.

There is a vast family of solutions that represents drops of different sizes separated by dry regions of different lengths. Thiele and Knobloch (2004) describe and illustrate a construction that generates all such solutions. An example for the hierarchical structure that such families form is given in Figure 14.

All of these solutions are nominally linearly stable, i.e. no coarsening can occur, since drops separated by truly dry regions do not interact if no non-hydrodynamic interaction is included. In the formulation without disjoining pressure the solutions with zero microscopic contact angle are energetically favored. However, the inclusion of a disjoining pressure selects a certain contact angle and removes the degeneracy. In Section 5 we discuss how the solution landscape collapses once the substrate is inclined.

In most works, like for instance, Oron and Rosenau (1992) and Thiele and Knobloch (2004), the temporal evolution of the pattern is restricted by rupture, i.e. it is limited to the short-time evolution, leaving the long-time limit un-investigated. In Bestehorn et al. (2003) a precursor film is stabilized by introducing a van der Waals term as disjoining pressure. The model then consists of the three-dimensional version of equation (3.1), the additional pressure equation (2.52), and the first term of equation (2.51). Linear stability analysis, construction of nonlinear steady solutions, as well as three-dimensional time dependent numerical solutions reveal a rich scenario of possible structures.

Using Maxwell-type constructions one can calculate the existence regions of drops, holes, or fronts with respect to the applied substrate temperature and the mean film thickness. On very thin films drops should always evolve at onset whereas for thicker films, the formation of holes is predicted. Drops or at least one big drop on a rather thick film was found in the experiment by VanHook et al. (1997) for an air layer, instead of holes for a helium gas layer above a silicone film. The thermal properties of the gas layer influence the Biot number and also the Marangoni number. It seems possible that the helium experiment was performed closer to threshold than the one with the air layer. According to the stability analysis in Bestehorn et al. (2003) this would explain the patterns observed in this experiment.

3.4   Coarsening

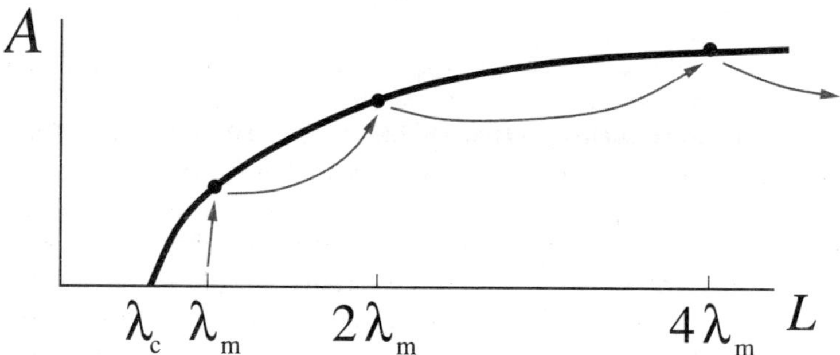

Figure 15. Schematic of the time evolution of a thin film. In the short-time evolution the flat film ruptures and the system approaches the large-amplitude branch. In the long-time evolution coarsening sets in and the system moves towards structures of larger length scales.

Up to here our presentation has mainly focused on the short-time evolution, i.e. the initial structuring process leading to the evolution of large amplitude structures from the initial flat film state. The time and length scales of the short-time evolution are determined by the linear modes of the flat film as sketched in Figure 15. In a first approximation the length scale is given by the wave length of the fastest growing linear mode [Equation (3.9)] and the time scale by $\tau_m = 1/\beta_m$ [Equation (3.9)]. A nonlinear rupture time can also be determined (Sharma and Jameel, 1993), but is normally of the same order of magnitude as the linear one.

However, all the large-amplitude periodic drops and holes are only stable when the analysis of their stability is restricted to instability modes of the same period. They are all unstable to modes of larger period, so-called 'coarsening modes'. For large times, a coarsening process then leads in consecutive steps to the appearance of larger and larger structures as sketched in Figure 15. Only a structure of system size is absolutely stable. An example of part of such a process is illustrated below by an energy-time plot (see below Figure 21, solid line). Note, that practically the coarsening process may be stopped by small substrate heterogeneities (see Section 4 below).

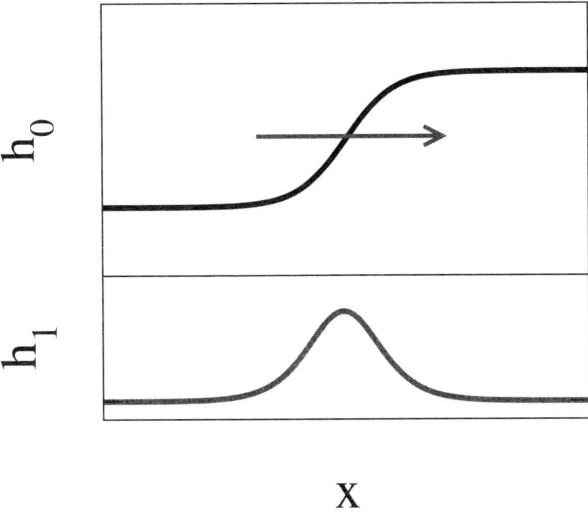

Figure 16. Sketch of the steady solution h_0 and the translational symmetry mode $h_1(x) = \partial_x h_0(x)$ for a liquid front.

In a system where the basic structure consists of drops, in the course of time droplets merge into larger droplets. Thereby the system moves along the low-energy branch of periodic solutions (Figure 10) towards states of lower energy. The pathway of the coarsening, i.e. the detailed properties of the active coarsening modes, can be understood from an analysis of the linear stability of the steady solutions on the low-energy branch. The important modes are related to the symmetry modes of the system. In general, each continuous symmetry is connected with a marginally stable symmetry mode, i.e. with a linear mode $h_1(x)$ that fulfills equation (3.14) with $\beta = 0$. For instance, situations involving a homogeneous horizontal substrate are translationally invariant. The related symmetry mode is the translation mode with $h_1(x) = \partial_x h_0(x)$ where h_0 is the steady solution. For a liquid front both are sketched in Figure 16.

To introduce the symmetry modes for a single droplet on a horizontal homogeneous substrate we assume that the left and the right side of the droplet are nearly decoupled allowing to see them as two individual fronts. Then we combine the symmetry modes

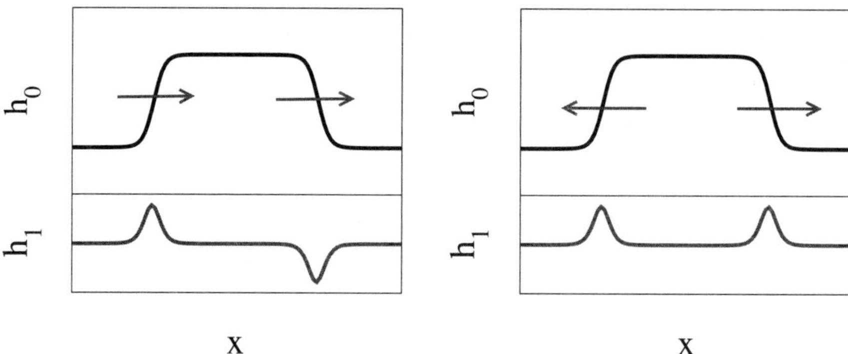

Figure 17. Symmetry modes for a single droplet on a horizontal substrate. Translation mode (left) and volume mode (right) are both obtained as a linear combination of translation modes for the two individual fronts.

of the two fronts. We obtain a translational mode when both fronts move in the same direction and a volume mode when they move in opposite directions (see Figure 17).

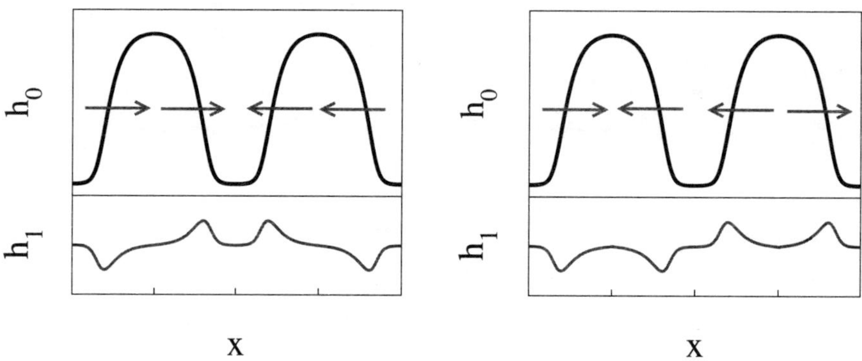

Figure 18. Coarsening modes corresponding to combinations of symmetry modes for individual droplets. Shown are translation mode (left) and mass transfer mode (right).

The two symmetry modes of a single droplet can be combined in the same spirit as above to obtain the two relevant coarsening modes for a pair of droplets. One mode results as a combination of a leftwards and a rightwards translation of the two respective drops (left part of Figure 18). This leads to coarsening by translation. The other mode combines a negative (inward) and a positive (outward) volume mode, i.e. one droplet grows at the cost of the other (right part of Figure 18). This is the mass transfer mode of coarsening. Material may be transported through the precursor film or through evaporation/condensation -assuming the transport equation (3.1) excludes the latter.

For unstable heated films a van der Waals interaction may stabilize a precursor film and avoid true rupture (Bestehorn et al., 2003). A numerical study then reveals the long-time behavior of such a system. The overall coarsening behaviour is well characterized by a scaling law $k \sim t^{-\nu}$ for the dependence of the typical length-scale $1/k$ on time t. An independent scaling factor with respect to the Hamaker constant close to threshold was found. First results give for the scaling exponent as $\nu = 0.21 \pm 0.01$. This indicates that the coarsening is slower than in spinodal decomposition, where $\beta = 1/3$ as given by the Lifshitz-Slyozov-Wagner theory (see for example Langer (1992)). The inclusion of hydrodynamic effects in the description of spinodal decomposition gives even larger exponents for the long time limit (in two dimensions $\beta = 1/2$ viscosity controlled, $\beta = 2/3$ inertia controlled (Podariu et al., 2000)). The scaling behavior is in general not yet well studied and we are still far from a detailed understanding (but see Mitlin (2001); Bestehorn et al. (2003); Glaser and Witelski (2003); Pismen and Pomeau (2004)). This is also true for two-layer films where a variety of scaling exponents is reported Merkt et al. (2005).

4 Horizontal Inhomogeneous Substrate

Several recent experiments (Karim et al., 1998; Rehse et al., 2001; Rockford et al., 1999; Sehgal et al., 2002) involve dewetting of thin films on inhomogeneous substrates. Most aim at arranging soft matter in a regular manner as determined by the physically and/or chemically patterned substrates. In nearly all experiments the (strong) heterogeneity imposes itself on the dewetting film if the length scale of the pattern is similar to the intrinsic scale of dewetting. This corresponds to theoretical results of a variety of groups (Lenz and Lipowsky, 1998; Bauer et al., 1999; Bauer and Dietrich, 2000; Kargupta et al., 2000, 2001; Kargupta and Sharma, 2001; Brinkmann and Lipowsky, 2002) based on a static approach using energy minimization or a dynamic approach using a long wave equation like equation (3.1) but with strong stepwise wettability contrasts (for a discussion see Thiele (2003a)). Deposited liquid volume, chemical potential or the size of the heterogeneous patches are used as control parameters to derive morphological phase diagrams. However, special care has to be taken using stepwise wettability patterns in dynamical studies because equation (3.1) is a long wave equation. In its derivation it is assumed that *all* relevant length scales parallel to the substrate are large as compared to the film thickness. This is not the case for a stepwise wettability contrast.

To understand the influence of a heterogeneous substrate in detail it is convenient to regard dewetting on a smoothly patterned substrate using the wettability contrast as a control parameter (Thiele et al., 2003). A film on a striped substrate is modeled using equations (3.1) and (2.50) replacing the constant κ by the heterogeneity

$$\kappa(x) = \kappa_0 \left[1 + \epsilon \cos \left(\frac{2\pi x}{P_{het}} \right) \right] \tag{4.1}$$

thereby modulating the wettability of the substrate sinusoidally. κ_0 is then absorbed into the scaling. The parameters ϵ and P_{het} correspond to the maximal wettability contrast and the period of the stripe-like heterogeneities.

Again continuation is used to determine all the stable and unstable steady solutions. Solutions determined for the homogeneous substrate or analytic solutions obtained for small ϵ are used as starting solutions. Then ϵ is increased to get the basic bifurcation diagram for the transition from homogeneous to inhomogeneous substrates and as a consequence also the characteristics of the transition between coarsening and pinning (ideal templating) and its dependence on heterogeneity strength, heterogeneity period and film thickness. Taking two periods of the heterogeneity as the system size one obtains, for instance, the bifurcation diagram given in Figure 19.

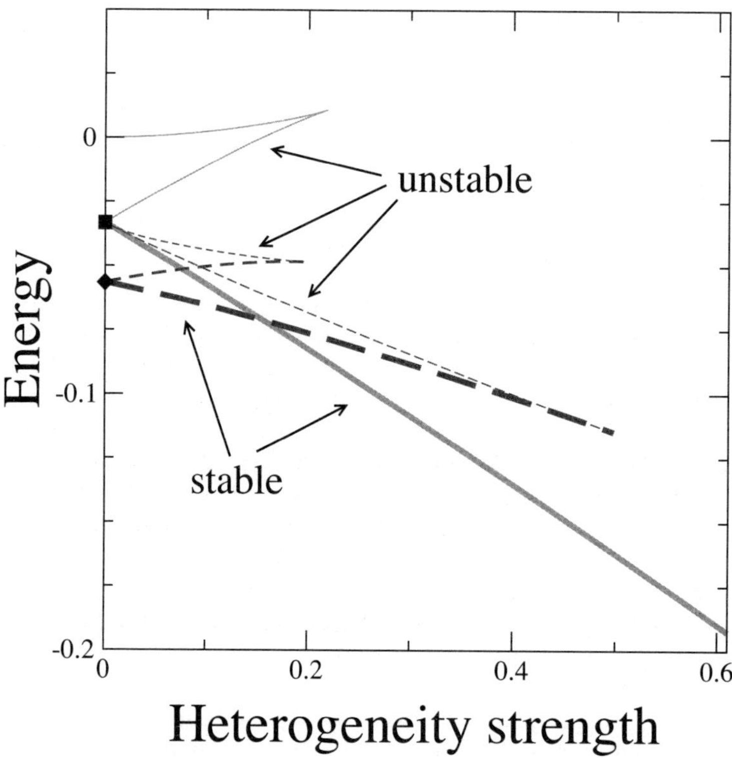

Figure 19. Energy of steady solutions to equation (3.3) with (4.1) for system size $L = P_{het}$ and $L = 2P_{het}$ versus the wettability contrast ϵ. Square and diamond denote the periodic solutions in the homogeneous case. Solid curves correspond to $L = P_{het}$ and dashed curves to $L = 2P_{het}$. Stable (unstable) solutions are marked by thick (thin) lines. The disjoining pressure is given by (2.50) and the parameters are $\bar{h} = 2.5$, $P_{het} = 50$ and $G = 0.1$ (for details see Thiele et al. (2003)).

We start the analysis with the shortest system of size $L = P_{het}$. Switching on the heterogeneity implies that the flat film is no longer a solution of equation (3.3) with (4.1). The flat film solution is replaced by a periodic solution that can be given analytically in the limit of weak heterogeneity $\epsilon \ll 1$. Writing the heterogeneity as $\epsilon \cos(kx) = \epsilon/2(e^{ikx} - c.c.)$ with $k = 2\pi/P$ where c.c. stands for complex conjugate, the volume conserving ansatz for the film thickness

$$h(x) = \bar{h} + \frac{\delta}{2}(e^{i\tilde{k}x} - \text{c.c.}) \tag{4.2}$$

is inserted into equation (3.3). This gives $\delta \sim O(\epsilon)$ and by neglecting terms of higher order in ϵ solutions with $\tilde{k} = k$

$$\delta = \frac{M(\bar{h})}{k_c^2 - k^2}\,\epsilon \tag{4.3}$$

where M is the κ-dependent part of the disjoining pressure. The condition $\delta \ll 1$ is only fulfilled if P is well separated from λ_c. For equation (2.50), $M(\bar{h})$ is always positive, hence for $k_c > k$ ($\lambda_c < P$) the solution is modulated in phase with the heterogeneity, i.e., the film is thicker on the less wettable parts of the substrate. These small amplitude solutions are linearly unstable like the flat film of the corresponding thickness without heterogeneity. However, if the critical period is larger than the period of the heterogeneity ($\lambda_c > P$) this solution has a phase shift of π with respect to the heterogeneity, i.e., the film is thinner on the less wettable parts of the substrate. This is also the case if the flat film of thickness $\bar{h}$ is linearly stable because there $k_c^2 < 0$. These small amplitude solutions are linearly stable.

Restricting our attention to linearly unstable film thicknesses we use the weakly modulated solution equations (4.2) and (4.3) as starting solutions for the continuation procedure (Doedel et al., 1997) and compute steady solutions with period $L = P_{het}$ as ϵ is increased. The flat film and periodic solution at $\epsilon = 0$ give rise to one and two solution branches, respectively. Figure 19 shows the corresponding bifurcation diagram. The branch emerging from the flat film solution is for small ϵ well approximated by the analytical result equation (4.3). The solutions on this branch are indeed in phase with the heterogeneity whereas the branch of lowest energy (thick solid line emerging from the square at $\epsilon = 0$) that corresponds to large amplitude solutions possesses a phase shift, as one would expect from physical considerations: the drops concentrate on the more wettable patches. The middle branch is in phase with the heterogeneity and terminates together with the small amplitude branch in a saddle-node bifurcation. In the small system with $L = P_{het}$, the entire thick solid branch in Figure 19 is linearly stable whereas the other two solid branches are linearly unstable.

New solutions appear as we double the system size to $L = 2\,P_{het}$. Besides the already discussed solutions with period P_{het}, there exist also solutions with period $2P_{het}$ (dashed lines in Figure 19). At $\epsilon = 0$ they are denoted by a diamond.

The solutions discussed above for system size $L = P_{het}$ have a period smaller than the system size and they may be unstable to coarsening. For $\epsilon = 0$ this is known from Section 3.4. Let us focus on the solutions on the thick solid branch that we will call 'pinned solutions' because they image the heterogeneity in an ideal way. For $\epsilon = 0$ a pinned solution possesses two positive eigenvalues corresponding to the coarsening modes of translation and mass transfer discussed above (Section 3.4). As ϵ departs from zero the heterogeneity counteracts the coarsening and the two modes are subsequently stabilized. For the present choice of parameters in Figure 19 the translational mode becomes stable at smaller ϵ values than the mass transfer mode implying that the mass transfer is the dominant coarsening process. As ϵ increases, both eigenvalues become negative, implying the linear stability of the pinned solution for larger ϵ. At the two points where the eigenvalues cross zero, period doubling bifurcations occur where steady solutions of period $2P_{het}$ emerge from the thick solid line. The two bifurcations are both subcritical hence the emerging solutions inherit the respective instability of the unstable solutions at smaller ϵ (cp. Guckenheimer and Holmes (1993)).

The four branches of solutions of period $2P_{het}$ shown in Figure 19 have the following linear stability characteristics. The upper thin dashed branch carries two positive eigenvalues and is itself a saddle in function space that divides evolutions by mass transfer towards the lower thin dashed branch and the branch of pinned solutions. The lower thin dashed branch still has one positive eigenvalue that leads to a shift of the pattern towards solutions of the thick dashed branch. The upper dashed branch emerging from the diamond has one positive eigenvalue and is a saddle that divides evolutions by translation of two droplets towards the thick dashed branch and the branch of pinned solutions. The entire thick dashed branch is linearly stable and represents the coarse solution competing with the pinned pattern.

From the results on the stability of the pinned and the coarse solution one can generally state that for a wettability contrast ϵ below a first threshold value ϵ_1 the coarse solution is the only stable once and also has the lowest energy. For $\epsilon > \epsilon_1$ one has multistability between the pinned and the coarse solution. As a consequence initial conditions and noise become very important for the evolution of the dewetting process. At $\epsilon = \epsilon_2$ the energies of both linearly stable solutions are equal. Above this value the pinned pattern has lowest energy. Finally at the threshold ϵ_3 the coarse solution ceases to exist and for $\epsilon > \epsilon_3$ the pinned solution is the only possible one.

Repeating the above analysis for the full range of P_{het} results in the morphological phase diagram Figure 20 where we indicate the relation of the different borders with the above defined ϵ_i's. At low values of ϵP_{het} coarsening prevails while for large values the pattern is pinned to the heterogeneity. At intermediate values (shaded area) the initial condition selects the asymptotic state due to multistability. In particular, it is not possible to pin a pattern to a heterogeneity that is much smaller than the critical wavelength λ_c for the surface instability of the corresponding flat film on the homogeneous substrate. Similar phase diagrams are also obtained for stepwise wettability contrast (Kargupta and Sharma, 2001, 2002). This suggests that the actual functional form of the heterogeneity is much less important than its length scale and strength.

The bifurcation and stability analysis of the possible liquid ridge solutions reveal parameter ranges where pinning or coarsening ultimately prevail, but also allows to establish the existence of a large hysteresis between pinned and coarse solutions, i.e. a large range where both morphologies correspond to local minima of the energy. This characterizes the pinning-coarsening transition as a first-order phase transition. In the resulting metastable range, the selected pattern depends sensitively on the initial conditions and potential finite perturbations (noise) in the system as illustrated with numerical integrations in time in (Thiele et al., 2003). Figure 21 shows that, compared to the homogeneous system, a weak heterogeneity slows down the onset of coarsening but accelerates the coarsening in the nonlinear regime. Recently, Grün et al. (2005) derived a stochastic evolution equation for a thin film on a heterogeneous substrate from the Navier-Stokes equations using long wave approximation and Fokker-Planck-type arguments.

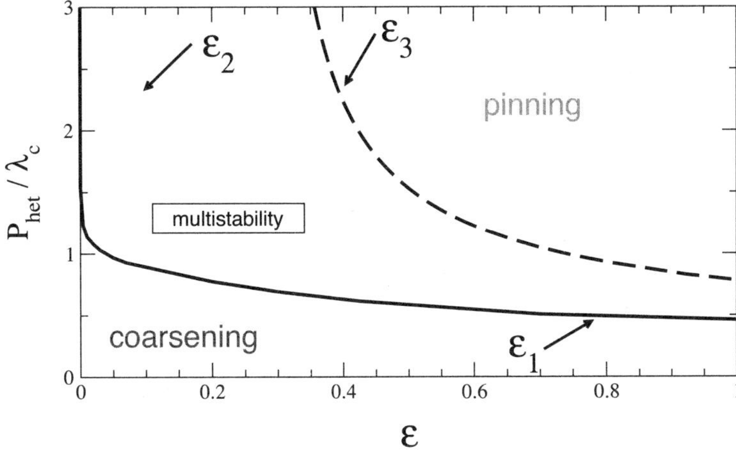

Figure 20. Morphological phase diagram of templating showing parameter regions in the $(\epsilon, P_{het}/\lambda_c)$ plane with different behavior of the thin film on a heterogeneous substrate. The shaded area separates the parameter region of coarsening from the one of pinning. Inside the shaded band multistability is present with the pinned pattern having the minimal energy inside the dark shaded area. The threshold values ϵ_i are indicated. Parameters are $\bar{h} = 2.5$ and $G = 0.1$ (for details see Thiele et al. (2003)).

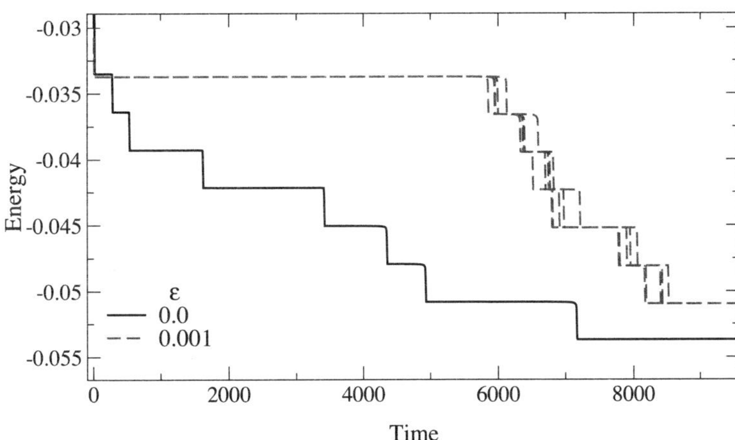

Figure 21. Energy for the long-time evolution for the homogeneous substrate (solid lines) and a slightly heterogeneous substrate (dashed lines, $\epsilon = 0.001$). The various dashed lines stand for evolutions with different noise but the same amplitude, 0.001. The remaining parameters are $\bar{h} = 2.5$, $P_{het} = 50$ and $G = 0.1$ (see Thiele et al. (2003)).

5 Inclined Homogeneous Substrate

5.1 General Analysis

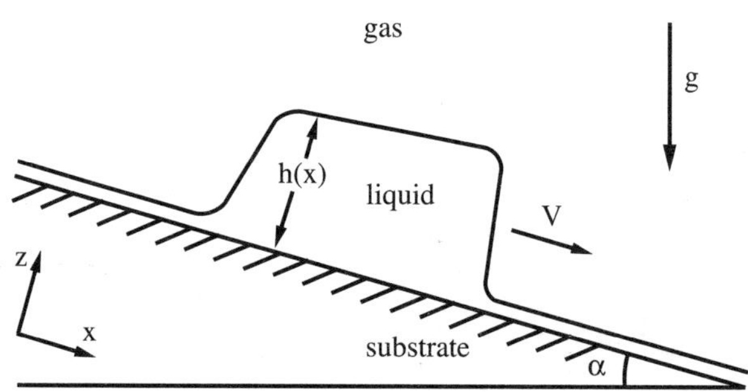

Figure 22. Sketch of a sliding drop on an inclined homogeneous substrate.

For a film on an inclined substrate we are interested in stationary moving solutions for the film thickness profile $h(x,t)$. We use the comoving coordinate system $x' = x - vt$, where v is the dimensionless velocity. Replacing in equation (2.47) the time derivative $\partial_t h$ by $-v\partial_{x'} h$ and integrating yields after dropping the prime

$$0 = Q(h)(\partial_{xxx} h - \partial_{hh} f \, \partial_x h) + \chi(h) - v\,h - C_0. \tag{5.1}$$

Note, that v is the phase velocity of the profile, not the mean velocity of the liquid (cf. surface waves). In contrast to the reflection symmetric problem of a film on a horizontal homogeneous substrate (Section 3), here we cannot set the integration constant C_0 to zero. Consequently one cannot integrate once more. Writing equation (5.1) in the form $C_0 = (\Gamma - vh)$ tells that for all stationary solutions the flow in the comoving frame $\Gamma - vh$ is constant but not the flow in the laboratory frame Γ. The choice of the constant of integration,

$$C_0 = (\Gamma_0 - vh_0) = -\chi(h_0) + v\,h_0, \tag{5.2}$$

introduces a flat film solution of thickness h_0. This corresponds to prescribing the liquid volume (see Scheid et al. (2005) when the flow rate is prescribed).

Besides the flat film solution given by the choice of h_0, there may exist other film thicknesses, h_i, that give the same flow in the comoving frame, $C_0 = \Gamma_i - vh_i = \Gamma_0 - vh_0$. These flat film solutions correspond to the fixed points of equation (5.1) seen as a dynamical system. Taking for instance, $\chi(h) = \alpha h^3$, there exist for a given h_0 two more flat film solutions

$$h_{1,2} = h_0 \left(-\frac{1}{2} \pm \sqrt{\frac{v}{\alpha h_0^2} - \frac{3}{4}} \right). \tag{5.3}$$

Because the film thickness has to be positive everywhere one has to choose the positive sign in (5.3). This solution is positive for $v/\alpha h_0^2 > 1$, i.e. a second flat film solution exists (called 'conjugate solution' in, for instance, Nguyen and Balakotaiah (2000)). Note, that the flat film solutions do not depend on the disjoining pressure $\Pi(h)$. However, they are entirely determined by $\Pi(h)$ in the limiting case $\alpha = 0$ (see equation (3.4). The consequences are discussed by Thiele et al. (2001c).

A linear stability analysis of the flat film yields the dispersion relation

$$\beta = -Q(h_0)\, k^2 \left(k^2 + \partial_{hh} f(h_0)\right) - i\, \partial_h \chi(h_0) k. \tag{5.4}$$

The real and imaginary parts of $\beta(k)$ give the respective growth rate and downwards phase velocity of the mode with wavenumber k.

In contrast to the existence of flat film solutions, their stability does not depend on the dynamical aspect of the problem (inclination angle and velocity), but only on $\partial_{hh} f(h_0)$ as for a flat film on a horizontal substrate (Section 3.1). In the limit of vanishing influence of the disjoining pressure this corresponds to the fact that for a pure Stokes flow, i.e. without the disjoining pressure, the flat film is linearly stable for all inclination angles. All linear modes propagate downwards with the velocity $v = -\mathrm{Im}\beta/k$ corresponding to the fluid velocity at the surface of the unperturbed flat film.

Consider a flat film in the linearly unstable thickness range. The will start to evolve in time. This may lead to stationary film profiles of finite amplitude, i.e. to sliding drops or nonlinear surface waves. They can be determined as periodic solutions of equation (5.1) using continuation methods (Doedel et al., 1997) starting from small amplitude analytic solutions.

5.2 Isothermal Partially Wetting Case

Experiments on liquid films or droplets moving down an inclined plate study, for instance, the formation of surface waves (Kapitza and Kapitza, 1949), localized structures and their interaction (Liu and Gollub, 1994), or shape transitions of sliding drops (Podgorski et al., 2001). Thereby, most studies on surface waves focus on structure formation caused by inertia measured by the Reynolds number and modeled by Benney-type equations (Benney, 1966; Joo et al., 1991; Scheid et al., 2005). However, because there the base flow and the 'structuring influences' have different orders in the smallness parameter (see Section 2.7), this type of description can normally not be applied to very thin films or to sliding drops.

Other studies that examine the evolution of falling sheets or ridges on an inclined 'dry' substrate encounter the problem of the moving three-phase contact line (Huppert, 1982; Silvi and Dussan, 1985) discussed above in Section 2.4. There the classical no-slip boundary condition at the liquid-solid interface has to be relaxed to permit movement of the contact line. This is normally done by introducing explicitly a very thin precursor film, or by allowing for slip near the contact line, or introducing an effective molecular interaction between the substrate and liquid into the hydrodynamic model (de Gennes, 1985; Dussan, 1979; Greenspan, 1978; Hocking, 1977; Huh and Scriven, 1971). Most work on moving liquid sheets and ridges uses one of the first two options. Both prescriptions avoid divergence problems at the contact line, but introduce ad-hoc parameters into the theory. These parameters, namely the slip length or the precursor film thickness, influence the base state profile and also the characteristics of transverse front instabilities (Bertozzi and Brenner, 1997; Hocking and Miksis, 1993; Kataoka and Troian, 1997; Spaid and Homsy, 1996). Moreover, the equilibrium and dynamic contact angles have to be fixed independently when introducing the slip condition (Greenspan, 1978; Hocking, 1990; Moyle et al., 1999). In contrast, in the absence of motion the precursor film models require that the contact angle be zero, although once the film is in motion the dynamic contact angle depends on the velocity of the advancing front.

Furthermore, most works (Hocking, 1990; Huppert, 1982; Troian et al., 1989) on liquid sheets or ridges flowing down an inclined plane analyze separately the three regions (1) upstream end, (2) central part and (3) downstream end of the ridge. Similarity solutions in the regions (1), (2) and (3) are matched together. The power-law dependence on time in these models is due to the fact that the situations studied are a superposition of spreading and sliding going on forever. However, recent experiments studied stationary drops that slide down a plate without changing their shape (Podgorski, 2000).

The explicit introduction of molecular interactions between the film surface and the substrate into the hydrodynamic formalism by means of a disjoining pressure (Section 2.4) can resolve the divergence problem at the moving contact line (Eres et al., 2000; Bestehorn and Neuffer, 2001; Thiele et al., 2001b, 2002a). Depending on the particular problem treated, this disjoining pressure may incorporate long-range van der Waals and/or various types of short-range interaction terms Teletzke et al. (1988); Hunter (1992); Israelachvili (1992). Because these interactions are essential for the process of dewetting, studies of dewetting of a thin liquid film on a substrate generally involve a disjoining pressure (see Section 3). This allows to connect the normally well separated fields of pattern formation on horizontal and inclined substrates.

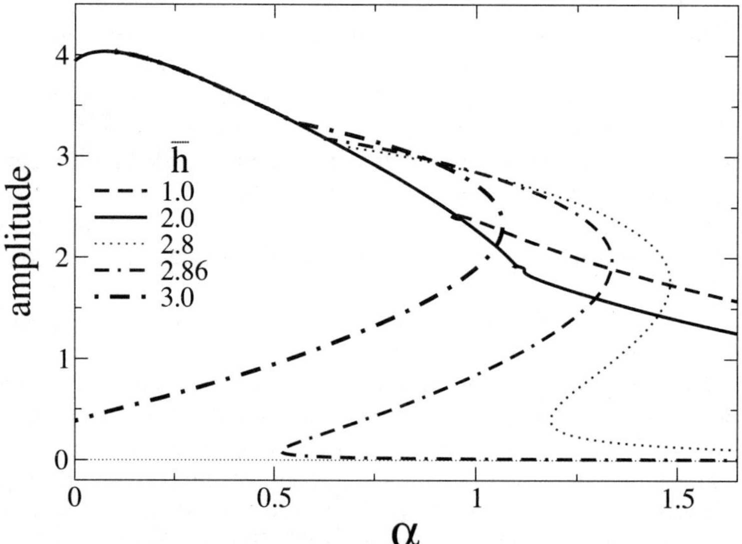

Figure 23. Characterization of stationary travelling periodic solutions. The figure shows the dependence of the amplitude on period, L, for different mean film thicknesses, $\bar{h}$, as given in the legend. $G = 0.1$ and $\alpha = 0.1$ (for details see Thiele et al. (2001b)).

However, only a few studies have adopted such an approach despite the fact that it predicts all the ad-hoc parameters of the slip or precursor models (i.e., the static and dynamic contact angle, drop velocity, and the drop and precursor film thickness). The necessary input is a description of the wetting properties of the liquid in terms of the parameters characterizing the disjoining pressure.

In the following we present some basic results for sliding drops and surface waves as stationary solutions of equation (3.1). Specifically, we present solutions of equation (5.1) for different mean film thicknesses and inclination angles. In general one distinguishes large amplitude drop-like solutions found for small inclination angles and small amplitude surface waves found for large inclinations. The solution families are presented for different mean film thicknesses in Figure 23 in the form of an inclination-amplitude plot. There the upper (lower) branches stand for drop-like (surface wave) solutions. Note, that for small inclinations the drop-like solutions all converge to a 'universal' curve (solid line). Drops along this line have a velocity and amplitude that are independent of their volume.

On the 'universal' curve the film profile converges to a common shape: a flat drop with a capillary ridge at the advancing front, an upper plateau of arbitrary extension and a non-oscillatory receding front. The drops slide on a very thin precursor film (lower plateau). Only the length of the upper plateau depends on the mean film thickness. Choosing one of the two plateau thicknesses as h_0, the other plateau thickness is given by equation (5.3). The velocity of the flat drops is determined by the dynamic equilibrium between the overall forces of gravity acting on the liquid and viscous friction. The equilibrium does not depend on the liquid volume because both forces are proportional to the length of the flat drop.

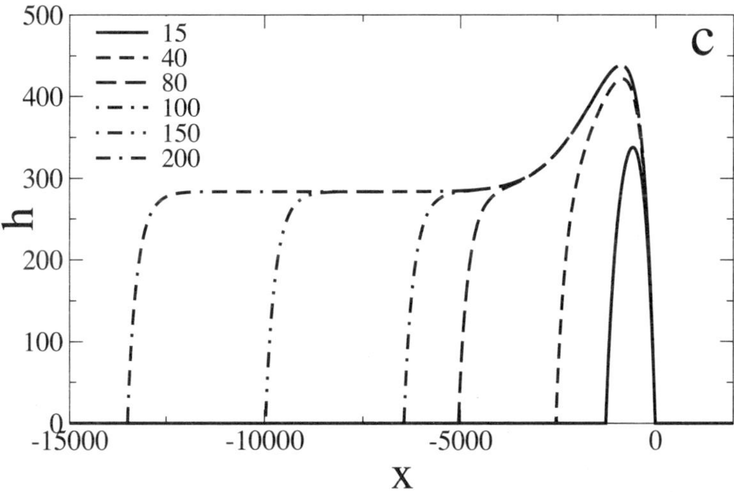

Figure 24. Profiles for different drop volume for a period of $L = 100000$, $G = 10^{-5}$ and $\alpha = 0.2$. The mean film thickness is increased (values given in legend, for details see Thiele et al. (2002a)).

Figure 24 shows the related transition, for fixed inclination angle and interaction parameters, from small (asymmetric) cup-like drops to large flat drops as the volume increases. The two-dimensional cup-like drops correspond to cross-sections of ridge-like solutions that are independent of the coordinate y transverse to the flow. The flat drops correspond to cross-sections of liquid sheets of finite longitudinal extent.

Using the models that incorporate a disjoining pressure allows to study the relation between the advancing and receding dynamic contact angles and the drop velocity. We define the contact angle as the absolute value of the slope of the profile at its inflection point. The drop has two dynamic contact angles, the advancing angle at its downstream front, θ_a, and the receding angle upstream, θ_r. The differences between these angles and the equilibrium or static contact angle, θ_e, obtained for $\alpha = 0$, are shown in Figures 9 and 10 of Thiele et al. (2001b), respectively.[10] The receding angle is always smaller than θ_e and in the velocity range studied here the absolute value of the difference between both angles increases linearly with increasing velocity. However, it was found that the difference between the advancing angle and θ_e changes non-monotonically with increasing velocity (Thiele et al., 2001b).

For a detailed analysis of the 'non-universal' part of the curves in Figure 23 we refer the reader to Thiele et al. (2001b, 2002a). Let us mention that the study of the non-universal solutions has revealed a hysteresis effect, when jumps between small and large amplitude solutions occur; both solutions exist for a certain range of inclination angles for some mean film thicknesses (cp. Figure 23). The transition between the two branches may be called a *dynamical wetting transition with hysteresis*. Such a dynamical wetting transition is now reasonably well understood (Blake and Ruschak, 1979; Podgorski, 2000; Pomeau, 2000; Ben Amar et al., 2001; Eggers, 2004): at the macro scale it occurs when the contact angle vanishes at finite velocity. However, as far as we know, no hysteresis has been predicted and/or observed although it does not seem to be excluded at all from first principles.

5.3 Heated Inclined Substrate

In Section 3.3 we already discussed the vast family of drop solutions that can be constructed for a heated horizontal substrate. Next we sketch how the solution landscape collapses once the substrate is inclined, i.e. using equation (2.47) with (2.52) and $\chi \neq 0$ (Thiele and Knobloch, 2004). On an inclined substrate the dry spots are replaced by regions of ultrathin film, and the array of drops slides downwards. This precursor film is not the usual static one but is dynamically produced by the 'first' drop. Its thickness depends on the dynamics of the system. This corresponds to the dependence of the lower plateau thickness on sliding velocity mentioned above in Section 5.2. In Thiele and Knobloch (2004) it is demonstrated that sliding trains of drops can be stable with respect to longitudinal perturbations, i.e. with respect to coarsening modes. Similar results have been obtained for convective Cahn-Hilliard models (Golovin et al., 2001).

Because of the enormous degeneracy of solutions in the horizontal case there is a very large number of slowly drifting drop-like states once the substrate is slightly inclined. In this regime one can locate intervals in the spatial period containing no stable simple trav-

---

[10] Note that the results for the contact angles in lubrication approximation have to be seen as corrections to the usually obtained proportionality $v \sim \theta^3$ (de Gennes, 1985) that is already absorbed into the scaling.

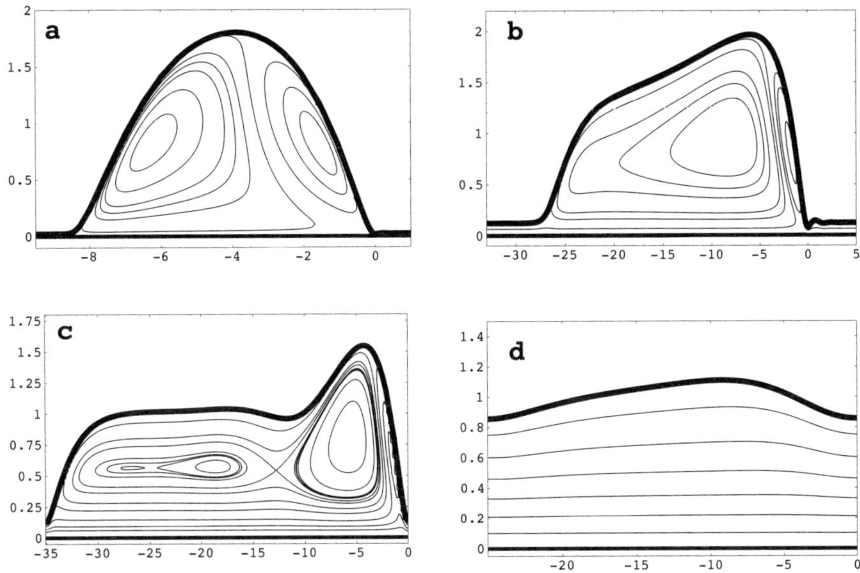

Figure 25. Comoving streamlines for different inclination angles and periods when Ma = 3.5, Bi = 0.5 and Bo = 0.5. (a) $\alpha = 0.08$, $L = 10$ and $v = 0.021$; (b) $\alpha = 0.08$, $L = 40$ and $v = 0.053$; (c) $\alpha = 0.215$, $L = 35$ and $v = 0.09$; (d) $\alpha = 0.215$, $L = 25$ and $v = 0.21$. The last panel shows a solution on the small amplitude surface wave branch formed when the primary bifurcation becomes supercritical (for details see Thiele and Knobloch (2004)).

eling wave solutions. Here one expects states with complex time-dependence, but perhaps more than that: the profusion of unstable states suggests that the system may wander among these states exhibiting very long transients even when stable states are available. At larger inclination the drift associated with finite inclination starts to dominate the dynamics, and the primary branches become supercritical with the resulting solutions resembling the traveling wavetrains familiar from the Kuramoto-Sivashinsky equation. As an illustration, Figure 25 presents a selection of stationary profiles for cap-like and flat drops and surface waves. The flow field in the comoving system is represented by streamlines.

The main conclusion is that the degeneracy of the horizontal case influences the inclined problem only for quite small values of the inclination angle where the dynamics is still Cahn-Hilliard-like. For larger inclinations the system behaves much more like the falling films under the influence of inertia studied, for example, by Benney (1966) and Joo et al. (1991), even though the theory does not retain inertial effects. Such films behave much like the Kuramoto-Sivashinsky equation, and Thiele and Knobloch (2004) have quantified the range of applicability of this equation to thin films on an inclined plane.

6 Transverse Instability

In the final section on systems described by the single film evolution equation (2.47), we focus on the transverse stability of liquid ridges on horizontal homogeneous, horizontal heterogeneous and inclined homogeneous substrates. Thereby we follow in part Thiele and Knobloch (2003) where the linear stability of liquid ridges on horizontal and inclined substrates is studied as a function of their volume and inclination. The aim is on the one hand to understand the transition between the various instability modes for liquid ridges on a horizontal substrate (Davis, 1980; Roy and Schwartz, 1999; Sekimoto et al., 1987) and on an inclined plane (Hocking, 1990; Hocking and Miksis, 1993), and on the other hand to relate these findings to results for falling semi-infinite sheets (Bertozzi and Brenner, 1997; de Bruyn, 1992; Eres et al., 2000; Huppert, 1982; Kalliadasis, 2000; Spaid and Homsy, 1996; Troian et al., 1989; Veretennikov et al., 1998; Ye and Chang, 1999).

In the latter case the leading front advances onto the 'dry' substrate and may initiate a fingering instability. It develops into an array of fingers advancing faster than the original front. A related transverse (or span-wise) instability occurs on a liquid front that advances as a result of a Marangoni flow induced by a longitudinal (i.e., stream-wise) temperature gradient (Bertozzi et al., 1998; Kataoka and Troian, 1998). In both cases the instabilities are due to differences in the mobility of the thinner and thicker parts of the capillary ridge at the advancing front.

6.1 Linear Stability Analysis of a Liquid Ridge

We have shown in the preceding sections that continuation is a very effective method to determine stable and unstable stationary solutions and their bifurcations by following them through parameter space using Newton's method (Doedel et al., 1991a,b). Here, we also apply it to study the linear stability of the constructed solutions with respect to transverse disturbances (Thiele and Knobloch, 2003). Now one has to continue simultaneously the stationary solution of equation (2.47), i.e. the profile $h_0(x)$ and the velocity v of the ridge, and the solution of equation (2.47) linearized around $h_0(x)$. Using the ansatz

$$h(x, y, t) = h_0(x) + \epsilon\, h_1(x)\, \exp(iky + \beta t) \tag{6.1}$$

to linearize equation (2.47), gives the linear eigenvalue problem for the growth rate β and disturbance h_1 as

$$\beta h_1(x) = \mathbf{S}[k, h_0(x)]\, h_1(x) \qquad \text{with} \qquad \mathbf{S}h_1 = \mathbf{N}_0 h_1 + k^2 \mathbf{N}_2 h_1 + k^4 \mathbf{N}_4 h_1 \tag{6.2}$$

with

$$
\begin{aligned}
\mathbf{N}_0 h_1 &= -\{Q_h h_1\,[(\partial_{xx}h_0 - \partial_h f)_x]\}_x\,, \\
&\quad -\{Q\,(\partial_{xx}h_1 - \partial_{hh}f h_1)_x\}_x \\
\mathbf{N}_2 h_1 &= \{Q\, h_{1x}\}_x + Q\,(\partial_{xx}h_1 - \partial_{hh}f h_1), \\
\mathbf{N}_4 h_1 &= -Q\, h_1.
\end{aligned}
\tag{6.3}
$$

The solution of equation (6.2) for the (complex) disturbance mode $h_1(x)$, transverse wavenumber k and the (complex) eigenvalue β has to be continued in parameter space together with the stationary solution $h_0(x)$. This is done using a three step procedure. First, h_0 and v are calculated by continuation starting from analytically known small amplitude solutions. The eigenvalue problem is then discretized in space and solved numerically. However, the necessary equidistant discretisation can be used in a small parameter range only. These restricted results are, in the third step, used as starting solutions for continuation of both h_0 and v, *and* of the solutions to the eigenvalue problem for any set of parameter values. The required extended system consists of 11 first order differential equations (3 for h_0 and 4 each for the real and imaginary parts of h_1). Furthermore, points of special interest such as the zeros or the maxima of the dispersion relation or the transition between real and complex eigenvalues can be followed through parameter space (Thiele and Knobloch, 2003).

6.2   Horizontal Homogeneous Substrate

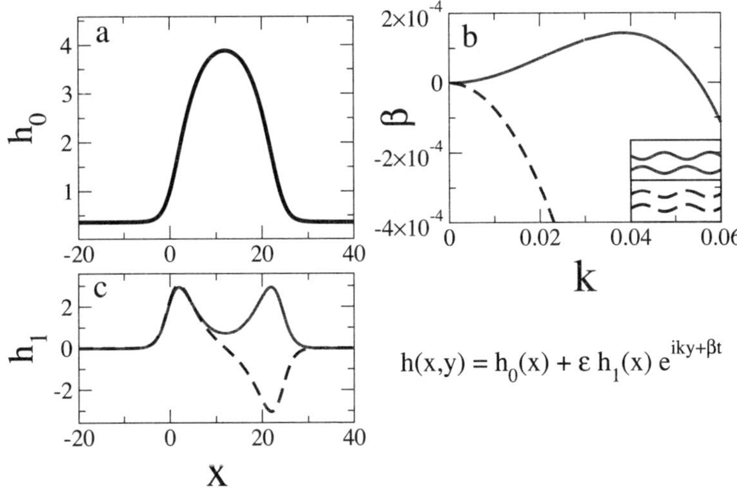

Figure 26. Overview for the transverse stability of a ridge on a horizontal substrate: (a) Steady ridge profile $h_0(x)$; (b) Dispersion relation $\beta(k)$ for the two transverse modes: the unstable varicose mode (solid) and the stable zigzag mode (dashed). Top views of the corresponding modes are sketched in the inset. (c) The eigenmodes $h_1(x)$ with line styles corresponding to (b). For details see Thiele and Knobloch (2003).

First we consider the homogeneous horizontal substrate. Recall that in this case the problem is variational implying that all eigenvalues β and eigenfunctions h_1 are real. This simplifies the above continuation procedure to that for 7 first order differential equations. An individual liquid ridge is always transversally unstable with respect to a varicose mode that emerges from the zero-eigenvalue at $k = 0$ connected to the volume mode for a single two-dimensional drop (cp. Section 3.4, especially Figure 17 (right)). It is reminiscent of the Rayleigh instability known from liquid jets (Chandrasekhar, 1992; Eggers, 1997). Figure 26 shows the cross section of such a ridge in (a), the dispersion relations in (b), and the relevant eigenmodes $h_i(x)$ in (c). The second important eigenmode is a zigzag mode emerging from the zero-eigenvalue at $k = 0$ representing the translation mode for a single two-dimensional drop (cp. Figure 17 (left)) It is always linearly stable on the horizontal substrate but becomes important for inclined substrates. Note, that the two modes correspond to transversally modulated volume (varicose) and translation (zigzag) modes of two-dimensional drops (see Figure 17 in section 3.4).

6.3 Horizontal Striped Substrate

On a striped substrate the wettability contrast not only stabilizes the system with respect to coarsening as detailed above in Section 4 but also stabilizes the transverse varicose instability (Gau et al., 1999; Thiele et al., 2003). However, having in mind the sinusoidal form of the heterogeneity (4.1) one cannot decouple individual ridges because their width is of the order of their distance. The interaction of the ridges influences also the transverse stability of an array of ridges. For details of the different mode types see Thiele et al. (2003). As an example, the varicose instability can take two forms: the instabilities of two neighboring ridges may be transversally in-phase or anti-phase. The in-phase combination is a periodic continuation of the varicose instability of an individual ridge and has the same dispersion relation. The growth rate goes to zero as the wavenumber approaches zero and the eigenmode at $k = 0$ corresponds to the volume mode. However, this is not the case for the anti-phase mode. For $k = 0$ it corresponds to the longitudinal coarsening mode by mass transfer between the two ridges as discussed in Section 3.4. For the zigzag mode one finds that the in-phase mode is stable as for an individual ridge, but the anti-phase mode has a small band of unstable wavenumbers around $k = 0$.

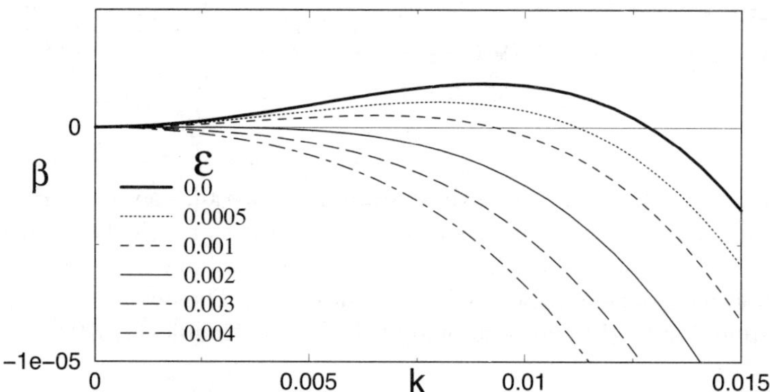

Figure 27. The largest eigenvalues β versus wavenumber k of transverse perturbations on a striped substrate. With increasing wettability contrast ϵ the range of unstable wave numbers becomes smaller, and finally vanishes (for details see Thiele et al. (2003)).

For the parameter values studied by Thiele et al. (2003) the in-phase varicose mode is the most important one for a transverse modulation of the ridges. Its stabilization with increasing wettability contrast is illustrated in Figure 27. On heterogeneous substrates all growth rates decrease monotonously with increasing strength of the heterogeneity. This leads first to a shrinkage of the band of unstable wavenumbers and then to the complete stabilization of the transverse instabilities. Already at small wettability contrast ϵ the length scale of the transverse instability is much larger than the longitudinal period. For $\epsilon = 0.001$, for instance, the fastest growing mode has a wavelength of about 1000 corresponding to twenty times P_{het}. The critical heterogeneity where the band of unstable varicose modes vanishes is only slightly larger than the value ϵ_1 where the longitudinal coarsening mode is stabilized (see Figure 20). The unstable zigzag mode stabilizes at smaller ϵ than the varicose modes.

Sliding ridge on inclined substrate. The physical situation changes once the substrate is inclined as a consequence of the broken symmetry $x \to -x$. As a result, the ridges become asymmetric and slide down the substrate (Thiele et al., 2001b)). When $\alpha = 0$ the variational structure of equation (2.47) implies that $v = C_0 = 0$. As a consequence, the equation is invariant under both translations in x and changes in volume (or $\bar{h}$). In particular, for each set of parameter values there is a two-parameter family of solutions. In contrast, when $\alpha \neq 0$ the stationary solutions are described by equation (5.1) with α, v, and C_0 all nonzero. The resulting equation is still invariant under translation but no longer under volume change. This is because a change in volume also changes the velocity v. As a result only the translational neutral mode remains, i.e. only one mode with zero growth rate exists at zero wavenumber, in contrast to the two modes for the horizontal substrate plotted in Figure 26. This implies that the two leading eigenmodes of the transverse stability problem, i.e. the equivalents of the varicose and zigzag modes for the inclined case, are either no longer independent at $k = 0$ or only one has a zero growth rate at $k = 0$. In fact we find that the first hypothesis is true: the two modes coincide at $k = 0$ forming the translational neutral mode, but are distinct whenever $k \neq 0$ and are mapped into one another by the transformation $k \to -k$.

The dispersion relation $\beta(k)$ undergoes dramatic changes when either increasing the ridge volume or the inclination of the substrate (Figure 28). Using as example the increase of the inclination, the different regimes are:

(i) For very small inclination, a varicose instability involving asymmetrically both edges of the ridge is still the dominant transverse instability mode. The zigzag mode is stable. Note that the mode changes along the dispersion curve. In the following we describe the mode at the maxima.

(ii) For slightly larger inclination the most unstable mode corresponds to an asymmetric zigzag type [Figure 28 (a)].

(iii) The dispersion curves of the zigzag and varicose modes couple resulting in complex modes for an intermediate wavenumber range [Figure 28 (b) and (c)]. Two maxima evolve.

(iv) The two modes decouple again and represent spatially decoupled instabilities of the advancing and the receding front having different growth rates and fastest growing wavenumbers ($k_{adv} < k_{rec}$). They are illustrated in a time simulation in Figure 29.

In the transition region between the instabilities (ii) - (iv), oscillatory instabilities are present in a certain wavenumber range. However, in the cases studied the oscillatory

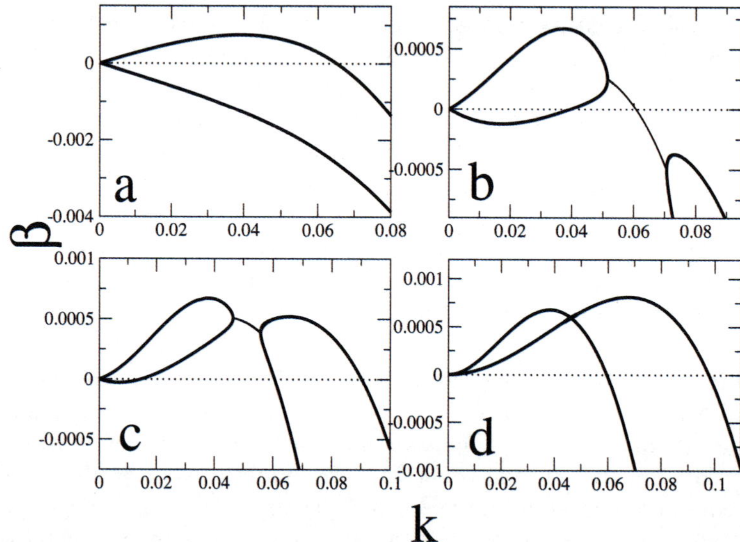

Figure 28. Sketch of the different dispersion relations obtained for increasing ridge volume or increasing inclination angle. Thick (thin) lines indicate real (complex) modes (for details see Thiele and Knobloch (2003)).

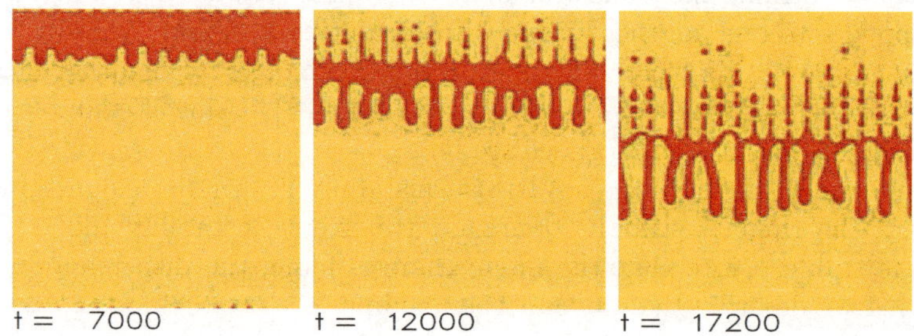

Figure 29. Time simulation of a sliding ridge. Both, the advancing and the receding front, get unstable to periodic transverse instabilities and fingering can be observed. $G = 0.2$, 512×512 points, $\alpha = 0.1$ and domain size 1900 (Thiele et al., 2002a).

modes were never dominant, and may therefore be seen only when the corresponding wavenumber is selected by the experimental apparatus.

The physical mechanism responsible for contact line instabilities can be studied using the method of energy analysis (Spaid and Homsy, 1996; Matar and Troian, 1997; Skotheim et al., 2003; Thiele and Knobloch, 2003). The growth rate β of an unstable mode is interpreted as an energy production rate and contributions to it from the individual terms of the linearized problem can be connected to underlying physical mechanisms (Spaid and Homsy, 1996). The decoupled instabilities of advancing and receding fronts are driven by gravity and by the destabilizing disjoining pressure, respectively (Thiele and Knobloch, 2003).

We close this Section with a speculation on the transverse instability found for dewetting. There holes often grow in a stable way (Redon et al., 1991; Reiter, 2001; Seemann et al., 2005), however, in a number of systems a transverse front instability of the receding dewetting front is observed. One can distinguish: (i) thickness modulations of the outward moving liquid rim around the growing hole (Brochard-Wyart and Redon, 1992; Masson et al., 2002; Meredith et al., 2000; Sharma and Reiter, 1996), (ii) development of relatively stable fingers that stay behind the outward moving rim (Herminghaus et al., 2000; Reiter, 1993; Thiele, 1998), and (iii) an emanation of a structured field of small droplets from the moving rim (Elbaum and Lipson, 1994; Kim et al., 1999; Reiter, 1993; Reiter and Sharma, 2001; Sharma and Reiter, 1996; van der Wielen et al., 2000). Not much is known on the exact conditions for the instability to occur and to have the type (i), (ii) or (iii). For the type (i) instability, it was proposed (Brochard-Wyart and Redon, 1992) that it is very similar to the Rayleigh instability of immobile rims as studied by Sekimoto et al. (1987). Type (ii) or (iii) instabilities were attributed (Reiter and Sharma, 2001) to a combination of a Rayleigh mechanism and dissipation due to slip (see also Thiele (1998)). The results of Thiele and Knobloch (2003) point to the destabilizing effect of the disjoining pressure, i.e. the effective interaction with the substrate that is responsible for the dewetting itself. Moreover, one may argue that the sequence of asymmetric varicose instability, asymmetric zigzag instability and decoupled front and back instability found when increasing the driving force, also gives a first hint on the mechanisms behind the change from instability type (i) to types (ii) and (iii) in dewetting. Also Figures 6 and 5 of Brochard-Wyart and Redon (1992) give a rough indication for a change from varicose to zigzag instability when increasing the driving force. A stronger zigzag instability leading to finger formation is also seen in Figure 14 of Reiter (1993).

7 Beyond the Single Evolution Equation

Finally, we introduce thin film systems that can not be described by a single evolution equation for the film thickness profile, i.e. by an evolution equation for a single conserved order parameter field. In the simplest case the description has then to be based on two coupled evolution equations. For systems involving a thin film with a free surface, the first equation describes the dynamics of the film thickness profile representing a conserved order parameter field. However, the second equation models a conserved *or* a non-conserved order parameter field.

We focus on two physical situations representing the respective two cases. In Section 7.1 we study ultrathin two-layer systems (Pototsky et al., 2004, 2005). There, the two mean interface heights are conserved, i.e. the two interface profiles correspond to the two conserved fields. In Section 7.2 we investigate chemically driven self-propelled running droplets (Thiele et al., 2004; John et al., 2005). There, the film thickness and the substrate adsorbate coverage correspond to the conserved and the non-conserved order parameter field, respectively.

Note, however, that we do not cover the extensive work on thin films with a layer of non-soluble surfactant that belongs to the first class, i.e. its dynamics is described by two coupled evolution equations of conserved order parameter fields (see Oron et al. (1997) and references therein).

7.1 Two-Layer Thin Films

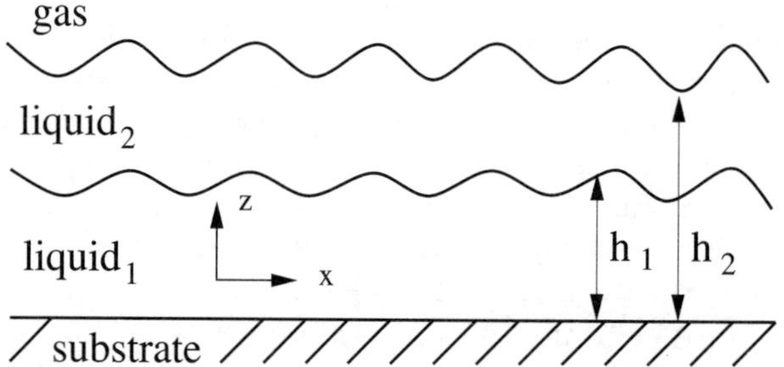

Figure 30. Geometry of the two-layer system.

Little is known about the behavior of two stacked ultrathin layers of soft matter on a solid substrate (see Figure 30). Such a two-layer film in an open geometry[11] allows for richer dynamics than the one-layer system because, both, the free liquid-liquid and the free liquid-gas interface evolve in a coupled way. The evolution is driven by the effective molecular interactions between *all* three interfaces separating the four material layers: substrate, liquid 1, liquid 2 and ambient gas. Although experimental studies investigate different aspects of dewetting for two-layer films, like interface instabilities or the growth of holes (David et al., 1998; Faldi et al., 1995; Lambooy et al., 1996; Morariu et al., 2003; Pan et al., 1997; Renger et al., 2000; Sferrazza et al., 1997, 1998), up to now the interface dynamics has not been studied in detail.

The most intricate question for the first stage of dewetting of a two-layer system is *which* interface will become unstable and *where* does the film rupture. This will determine the final morphology of the film. Experiments found roughening of the liquid-liquid interface (Sferrazza et al., 1998) or an instability of the liquid-gas interface (Faldi et al., 1995; Morariu et al., 2003). Holes that evolve solely in the upper layer were also studied (Lambooy et al., 1996; Pan et al., 1997).

---

[11]In the open geometry the system has two free interfaces. On the contrary, in a bounded geometry the two layers are enclosed by two solid substrates. Then the evolution of the single free interface can still be described by a single thin film equation (Merkt et al., 2005).

Evolution equations for the film thicknesses h_1 and h_2 can be obtained from the Navier-Stokes equations by employing long-wave approximation along the lines sketched above in Section 2 for a single layer. One obtains

$$\frac{\partial h_1}{\partial t} = \partial_x \left(Q_{11} \partial_x \frac{\delta F}{\delta h_1} + Q_{12} \partial_x \frac{\delta F}{\delta h_2} \right)$$
$$\frac{\partial h_2}{\partial t} = \partial_x \left(Q_{21} \partial_x \frac{\delta F}{\delta h_1} + Q_{22} \partial_x \frac{\delta F}{\delta h_2} \right), \tag{7.1}$$

where $\delta F / \delta h_i$ with $i = 1, 2$ denotes functional derivatives of the total energy of the system

$$F = \int [\rho_S + \rho_{VW}] \, dx \tag{7.2}$$

and Q_{ik} are the positive elements of the symmetric mobility matrix. The energy F contains the densities of the surface energy and of the energy for the van der Waals interaction (Pototsky et al., 2004). Inclusion of other interactions, like for instance, a stabilising short-range interaction, is straightforward (Pototsky et al., 2005). Related models were derived, for instance, assuming a lower liquid layer that is much thicker than the upper layer (Brochard-Wyart et al., 1993), and for two-layer systems with surfactants (and non-Newtonian behavior) (Zhang et al., 2003; Craster and Matar, 2000; Matar et al., 2002) or including evaporation (Danov et al., 1998a,b; Paunov et al., 1998). A two-layer system under the solely influence of long-range molecular interactions is studied by Bandyopadhyay et al. (2005).

The system (7.1) represents the most general form of coupled evolution equations for two conserved order parameter fields in a relaxational situation and is able to describe a broad variety of experimentally studied two-layer systems.

Studying equations (7.1) by means of linear analysis leads to two different types of unstable modes, namely, varicose and zigzag modes that also determine the non-linear evolution. In the model without stabilizing short-range interactions studied in Pototsky et al. (2004), the two modes lead to rupture at the substrate or at the liquid-liquid interface. Both modes are asymmetric since the deflection amplitudes of the two interfaces are normally different. The mobilities have no influence on the stability threshold, but determine mode type and the length and time scales of the dynamics.

Furthermore, the simultaneous action of the van der Waals forces between the three interfaces allows for dispersion relations with two maxima. An experimental system showing this unusual form of $\beta(k)$ can be realized with a substrate that is less polarisable than both layers of soft matter. If the two maxima are of equal height, two modes of different wavelength may evolve at the two interfaces, respectively. Different types of dispersion relations and snapshots of respective time evolutions are given in Figure 31. For more details see Pototsky et al. (2004).

The system presented here is strongly restricted by the sole inclusion of a destabilizing interaction. Every unstable evolution terminates with a rupture event. To be able to study different pathways of coarsening that may occur in the long-time evolution, one has to include a stabilizing short-range interaction. In this way thin precursor films are stabilized and a rich coarsening dynamics is found (Pototsky et al., 2005).

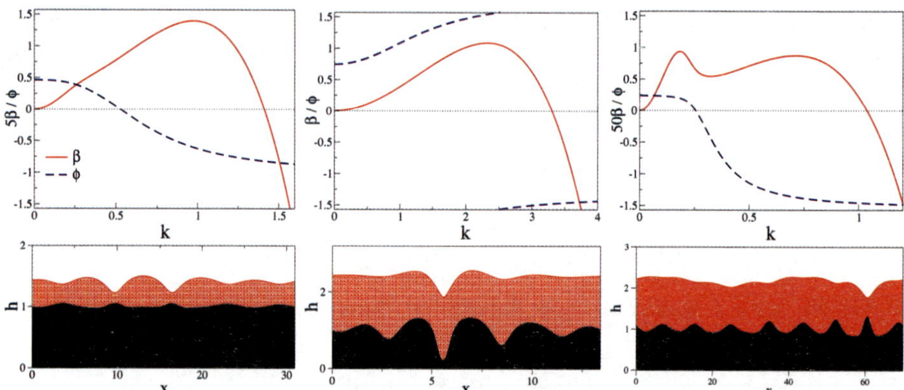

Figure 31. Growth rate γ (solid lines) and the mode type ϕ (dashed lines) of the leading eigenmode. (a) A varicose mode from the one-mode region at $d_1 = 30$, $d_2 = 47$ and $\sigma = \mu = 1$. (b) A zigzag mode from the one-mode region at $d_1 = 15$, $d_2 = 40$ and $\sigma = \mu = 1$. Panel (c) gives γ and ϕ for d_1 and d_2 as in (b) but for $\mu = 0.1$. For convenience we plot in (b) 10γ and in (a) 20γ. Snapshots from time evolutions of a two-layer film for a Si/PMMA/PS/air system at dimensionless times (in units of τ_{up}) are shown in the insets. (a) At $d = 1.4$ a varicose mode evolves leading to rupture of the upper layer at the liquid-liquid interface. The ratio of the time scales derived from upper and lower effective one-layer system is $\tau_{\mathrm{up}}/\tau_{\mathrm{low}} = 0.066$. (b) At $d = 2.4$ a zigzag mode evolves and rupture of the lower layer occurs at the substrate ($\tau_{\mathrm{up}}/\tau_{\mathrm{low}} = 34.98$). The domain lengths are 5 times the corresponding fastest unstable wave length and $\mu = \sigma = 1$ (for details see Pototsky et al. (2004)).

7.2 Chemically Driven Running Drops

It is generally known that drops can move in externally driven gradients, like for instance, in a temperature or chemical gradient (Brochard, 1989) or most simply on an inclined plate as discussed above in Section 5. However, drops may also move in an initially gradientless surrounding if they themselves change the surrounding and thereby produce a gradient that drives their motion.

Recent experiments found such droplets on solid substrates that are chemically changed by the droplets themselves. A driving wettability gradient is produced by an adsorption or desorption reaction at the substrate underneath the drop (Bain et al., 1994; Domingues Dos Santos and Ondarçuhu, 1995; Lee et al., 2002; Sumino et al., 2005a,b).

In these experiments, a small droplet of solvent is put on a partially wettable substrate. A chemical dissolved in the droplet reacts with the substrate resulting in the deposition of a less wettable coating. Likewise, it may also dissolve a wettable coating exposing the less wettable bare substrate. In both cases, the substrate below becomes less wettable than the substrate outside the droplet. Because of the radial symmetry it is still in an equilibrium position. However, it becomes more and more unstable. Eventually, the symmetry is broken by fluctuations and the drop starts to move, thereby changing the substrate and leaving a less wettable trail behind (see Figure 32). Similar phenomena can be seen in reactive (de)wetting (Zheng et al., 1998), in camphor boats (Hayashima et al., 2002), and in the migration of reactive islands in alloying (Schmid et al., 2000).

A simple theoretical argument (Brochard-Wyart and de Gennes, 1995) predicts a monotone increase of the droplet velocity with the droplet length and the reaction rate, in line with early experimental observations (Domingues Dos Santos and Ondarçuhu, 1995). Recent experiments (Lee et al., 2002) show also the opposite trend; the velocity decreases with increasing drop sizes and reaction rate.

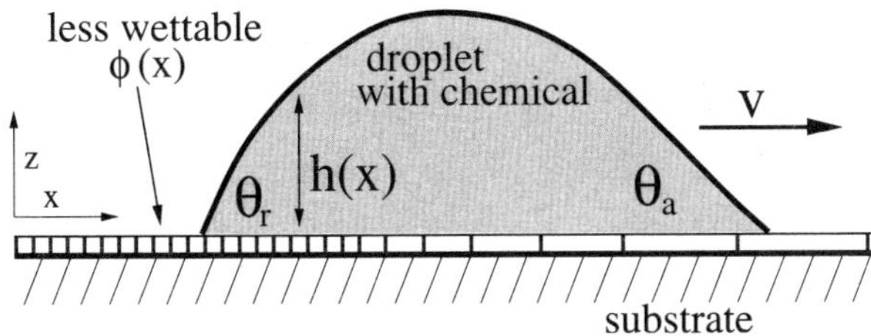

Figure 32. Sketch of a right moving droplet driven by a self-produced chemical gradient.

Recently we proposed and analyzed dynamical models for self-propelled running droplets (Thiele et al., 2004; John et al., 2005). Here we introduce a model that only accounts for an adsorption reaction underneath the droplets but not for recovery of the substrate. The model consists of coupled evolution equations describing the interdependent spatiotemporal dynamics of the film thickness h and of the concentration of an adsorbate ϕ that decreases the substrate wettability. The model equations in dimensionless form are

$$\partial_t h = -\nabla \left\{ h^3 \nabla \left[\Delta h - \partial_h f(h, \phi) \right] \right\} \tag{7.3}$$

$$\partial_t \phi = R(h, \phi) + d \Delta \phi. \tag{7.4}$$

The first equation describes the evolution of the film thickness profile. It is based on equation (2.47) with $\chi = 0$, and incorporates a disjoining pressure following equation (2.51) (with the positive sign). However, the short-range part of the disjoining pressure depends not only on h but changes linearly with the adsorbate concentration ϕ

$$\partial_h f(h, \phi) = -\frac{b}{h^3} + \left(1 + \frac{\phi}{g} \right) e^{-h}. \tag{7.5}$$

The scaled equilibrium contact angle θ_e is given by $\cos\theta_e = \hat{S} + 1$, where $\hat{S}$ is the dimensionless (negative) spreading coefficient (Sharma, 1993b) and $1/g$ defines the magnitude of the wettability gradient. With equation (7.5), $\hat{S} = b - 1 - \phi/g$ and $b \geq 0$, implying that θ_e increases with increasing ϕ, i.e. the coated substrate is less wettable.

The second equation (7.4) models the evolution of the chemical concentration of the adsorbate using a reaction-diffusion equation. The function $R(h, \phi)$ describes the reaction that changes the wettability of the substrate and the second term allows for a (usually small) diffusion of the chemical species along the substrate. The main results, however, are obtained without diffusion. As reaction term we choose

$$R(h, \phi) = r\Theta(h - h_0)\,(1 - \phi) \tag{7.6}$$

where r defines the time scale of the reaction. It is assumed that the reaction at the substrate occurs only underneath the droplet as modelled by the step function $\Theta(h - h_0)$, and that it saturates at a value $\phi_{max} = 1$. The value of h_0 is chosen slightly larger than the thickness of the precursor film.

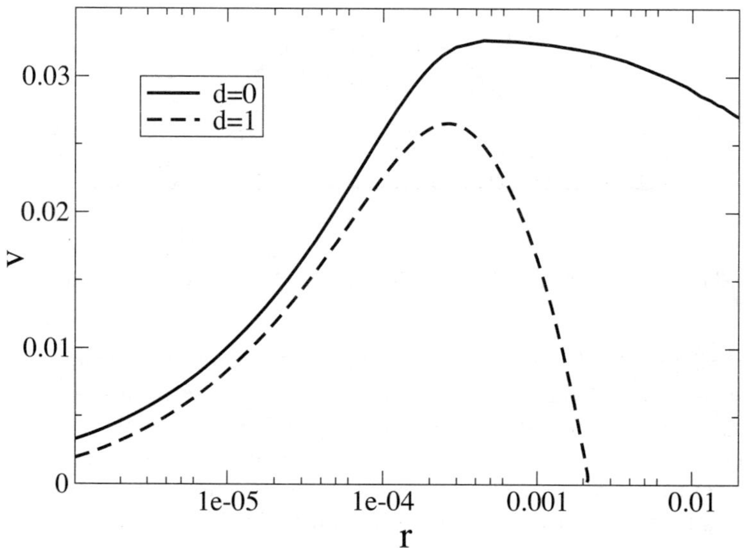

Figure 33. Characterization of running droplets stationary in a comoving frame. The figure shows the velocity v as a function of the reaction rate r (logarithmic scale) without ($d = 0$, solid line) and with ($d = 1.0$, dotted line) diffusion. The remaining parameters are $g = 1.0, d = 0.0, b = 0.5$ and the droplet volume is 30,000 (for details see Thiele et al. (2004)).

The model equations (7.3) and (7.4) are capable of reproducing the different experimentally found regimes. In particular, by varying the reaction rate or drop volume we identify two distinct regimes of running drops as shown in Figure 33. For small reaction rate (or droplet size), the chemical gradient in the drop is limited by the progress of reaction. In contrast, for a fast reaction (or large droplets), the chemical concentration at the receding end saturates at the maximum value. The velocity of reaction-limited droplets increases, while the velocity of saturated droplets decreases with increasing reaction rate or droplet size. Specifically, the velocity of the droplets may increase (Domingues Dos Santos and Ondarçuhu, 1995) or decrease (Lee et al., 2002) with their volume. If we take into account fast diffusion of the adsorbate along the substrate, the droplet velocity in the saturated regime decreases much faster and only steady sitting drops are found at large reaction rates.

In addition, we find that the dynamic contact angles at the advancing and receding edges of the droplets differ substantially. The differences between the static and the dynamic contact angles at the front and the rear are one order of magnitude smaller than the difference between the two static contact angles. This challenges the assumption of equal dynamic contact angles at the front and the rear that was used by de Gennes (1998) and Brochard-Wyart and de Gennes (1995) to develop a simple description of self-propelled running droplets. A simple quantitative theory should instead be based on the assumption that the respective dynamic contact angles equal the different static contact angles at the front and rear.

An extension of the model presented here describes also experiments (Sumino et al., 2005a,b) where the substrate recovers its original state behind the droplet through an adsorption from a surrounding medium (John et al., 2005). This allows, for instance, for a periodic droplet movement on circular or finite stripe-like substrates.

8 Outlook

We have given an overview of some recent developments in the description of pattern formation in thin liquid films. Although we have discussed the main developments in the field, the presentation of the derivation and analysis of thin film equations focused mainly on our own results. The outcome is a chapter that has in part introductory and in part review character.

Specifically, on the one hand, we have focused on the common mathematical framework behind all thin film systems involving a single layer of liquid. Thereby, we have emphasized the advantages of studying the transitions between different geometries, i.e. from homogeneous to inhomogeneous, or from horizontal to inclined substrates. On the other hand, we have introduced the physical questions posed by the individual systems detailing our contributions to them and their relation with the literature. We summarize here a selection of main results for single liquid layers:

- For the initial film rupture in the process of dewetting one has to distinguish nucleation-dominated and instability-dominated behavior for linearly unstable films.
- For heated films on a horizontal substrate we have discussed nucleation and drop solutions. We also showed that it is possible to construct all drop solutions separated by dry regions.
- Incorporating a disjoining pressure in the study of heated thin films on horizontal substrates has allowed us to study the long-time coarsening behavior of the evolving pattern.
- For dewetting on an inhomogeneous substrate we described a pinning-coarsening transition with a large range of multistability, implying a large hysteresis and strong dependence on initial conditions and noise.
- Studying sliding drops on an inclined homogeneous substrate by using a model that incorporates a disjoining pressure has allowed us to calculate from surface chemistry the usual ad-hoc parameters of models for moving contact lines.
- For heated films on slightly inclined plates we have described a complex transition from a Cahn-Hilliard-like to a Kuramoto-Sivashinsky-like dynamics occurring at small inclination angles.

- We have investigated the transverse instabilities of liquid ridges on homogeneous and striped horizontal substrates and for a sliding liquid ridge on an inclined substrate we have shown that the mode type of the instability changes with increasing inclination from a symmetric varicose mode (horizontal substrate) via an asymmetric varicose mode via an asymmetric zigzag mode to decoupled front and back modes.

Finally, we have shown two possible ways to extend the study of thin film systems beyond the case of a single evolution equation. Specifically, we have introduced models describing:

- The dynamics of a two-layer thin film that may follow different pathways of dewetting;
- Self-propelled droplets driven by a chemical reaction at the substrate.

These models describe some experiments from a large class of systems that have to be modeled by coupled evolution equations. Besides the cases studied here, the evolution of the film thickness can be accompanied by phase changes as observed, for example, for liquid crystals, block copolymers or polymer melts (Demirel and Jerome, 1999; Yerushalmi-Rozen et al., 1999; Knoll et al., 2002; Schlagowski et al., 2002) or by density variations as discussed in Sharma and Mittal (2002). To describe the observed phenomena involving complex fluids it will be necessary to derive coupled evolution equations for the film thickness and fields describing the involved inner degrees of freedom.

Bibliography

D. M. Anderson, G. B. McFadden, and A. A. Wheeler. Diffuse-interface methods in fluid mechanics. *Ann. Rev. Fluid Mech.*, 30:139–165, 1998.

C. D. Bain, G. D. Burnetthall, and R. R. Montgomerie. Rapid motion of liquid-drops. *Nature*, 372:414–415, 1994.

N. J. Balmforth, R. V. Craster, and R. Sassi. Dynamics of cooling viscoplastic domes. *J. Fluid Mech.*, 499:149–182, 2004.

D. Bandyopadhyay, R. Gulabani, and A. Sharma. Stability and dynamics of bilayers. *Ind. Eng. Chem. Res.*, 44:1259–1272, 2005.

S. G. Bankoff. Significant questions in thin liquid-film heat-transfer. *J. Heat Transf.-Trans. ASME*, 116:10–16, 1994.

C. Bauer and S. Dietrich. Phase diagram for morphological transitions of wetting films on chemically structured substrates. *Phys. Rev. E*, 61:1664–1669, 2000.

C. Bauer, S. Dietrich, and A. O. Parry. Morphological phase transitions of thin fluid films on chemically structured substrates. *Europhys. Lett.*, 47:474–480, 1999.

J. Becker, G. Grün, R. Seemann, H. Mantz, K. Jacobs, K. R. Mecke, and R. Blossey. Complex dewetting scenarios captured by thin-film models. *Nature Mat.*, 2:59–63, 2003.

M. Ben Amar, L. Cummings, and Y. Pomeau. Singular points of a moving contact line. *C R Acad. Sci. Ser. IIB*, 329:277–282, 2001.

H. Bénard. Les tourbillons cellulaires dans une nappe liquide. *Rev. Gén. Sci. Pures Appl.*, 11:1261–1271, 1900.

T. B. Benjamin. Wave formation in laminar flow down an inclined plane. *J. Fluid Mech.*, 2:554, 1957.

D. J. Benney. Long waves on liquid films. *J. Math. & Phys.*, 45:150–155, 1966.

A. L. Bertozzi and M. P. Brenner. Linear stability and transient growth in driven contact lines. *Phys. Fluids*, 9:530–539, 1997.

A. L. Bertozzi, A. Münch, X. Fanton, and A. M. Cazabat. Contact line stability and "undercompressive shocks" in driven thin film flow. *Phys. Rev. Lett.*, 81:5169–5173, 1998.

M. Bestehorn and K. Neuffer. Surface patterns of laterally extended thin liquid films in three dimensions. *Phys. Rev. Lett.*, 87:046101,1–4, 2001.

M. Bestehorn, A. Pototsky, and U. Thiele. 3D large scale Marangoni convection in liquid films. *Eur. Phys. J. B*, 33:457–467, 2003.

J. Bischof, D. Scherer, S. Herminghaus, and P. Leiderer. Dewetting modes of thin metallic films: Nucleation of holes and spinodal dewetting. *Phys. Rev. Lett.*, 77:1536–1539, 1996.

T. D. Blake and K. J. Ruschak. A maximum speed of wetting. *Nature*, 282:489–491, 1979.

W. Boos and A. Thess. Cascade of structures in long-wavelength Marangoni instability. *Phys. Fluids*, 11:1484–1494, 1999.

M. Brinkmann and R. Lipowsky. Wetting morphologies on substrates with striped surface domains. *J. Appl. Phys.*, 92:4296–4306, 2002.

F. Brochard. Motions of droplets on solid-surfaces induced by chemical or thermal-gradients. *Langmuir*, 5:432–438, 1989.

F. Brochard-Wyart and J. Daillant. Drying of solids wetted by thin liquid films. *Can. J. Phys.*, 68:1084–1088, 1989.

F. Brochard-Wyart and P.-G. de Gennes. Spontaneous motion of a reactive droplet. *C. R. Acad. Sci. Ser. II*, 321:285–288, 1995.

F. Brochard-Wyart, P. Martin, and C. Redon. Liquid/liquid dewetting. *Langmuir*, 9: 3682–3690, 1993.

F. Brochard-Wyart and C. Redon. Dynamics of liquid rim instabilities. *Langmuir*, 8: 2324–2329, 1992.

F. Brochard-Wyart, C. Redon, and C. Sykes. Dewetting of ultrathin liquid films. *C. R. Acad. Sci.*, 314 II:19–24, 1992.

J. P. Burelbach, S. G. Bankoff, and S. H. Davis. Nonlinear stability of evaporating/condensing liquid films. *J. Fluid Mech.*, 195:463–494, 1988.

J. P. Burelbach, S. G. Bankoff, and S. H. Davis. Steady thermocapillary flows of thin liquid layers. II. Experiment. *Phys. Fluids A*, 2:321–333, 1990.

J. W. Cahn. Phase separation by spinodal decomposition in isotropic systems. *J. Chem. Phys.*, 42:93–99, 1965.

J. W. Cahn and J. E. Hilliard. Free energy of a nonuniform system. 1. Interfacual free energy. *J. Chem. Phys.*, 28:258–267, 1958.

S. Chandrasekhar. *Hydrodynamic and Hydromagnetic Stability*. Clarendon Press, Oxford, 1992.

H.-C. Chang. Wave evolution on a falling film. *Ann. Rev. Fluid Mech.*, 26:103–136, 1994.

R. V. Craster and O. K. Matar. Surfactant transport on mucus films. *J. Fluid Mech.*, 425:235–258, 2000.

M. C. Cross and P. C. Hohenberg. Pattern formation out of equilibrium. *Rev. Mod. Phys.*, 65:851–1112, 1993.

K. D. Danov, V. N. Paunov, N. Alleborn, H. Raszillier, and F. Durst. Stability of evaporating two-layered liquid film in the presence of surfactant - I. The equations of lubrication approximation. *Chem. Eng. Sci.*, 53:2809–2822, 1998a.

K. D. Danov, V. N. Paunov, S. D. Stoyanov, N. Alleborn, H. Raszillier, and F. Durst. Stability of evaporating two-layered liquid film in the presence of surfactant - ii. linear analysis. *Chem. Eng. Sci.*, 53:2823–2837, 1998b.

M. O. David, G. Reiter, T. Sitthai, and J. Schultz. Deformation of a glassy polymer film by long-range intermolecular forces. *Langmuir*, 14:5667–5672, 1998.

S. H. Davis. Moving contact lines and rivulet instabilities. Part 1. The static rivulet. *J. Fluid Mech.*, 98:225–242, 1980.

S. H. Davis. Thermocapillary instabilities. *Ann. Rev. Fluid Mech.*, 19:403–435, 1987.

J. R. de Bruyn. Growth of fingers at a driven three–phase contact line. *Phys. Rev. A*, 46:R4500–R4503, 1992.

P.-G. de Gennes. Wetting: Statistics and dynamics. *Rev. Mod. Phys.*, 57:827–863, 1985.

P.-G. de Gennes. The dynamics of reactive wetting on solid surfaces. *Physica A*, 249: 196–205, 1998.

R. J. Deissler and A. Oron. Stable localized patterns in thin liquid films. *Phys. Rev. Lett.*, 68:2948–2951, 1992.

A. L. Demirel and B. Jerome. Restructuring-induced dewetting and re-entrant wetting of thin glassy films. *Europhys. Lett.*, 45:58–64, 1999.

B. V. Derjaguin, N. V. Churaev, and V. M. Muller. *Surface Forces*. Consultants Bureau, New York, 1987.

E. Doedel, H. B. Keller, and J. P. Kernevez. Numerical analysis and control of bifurcation problems (I) Bifurcation in finite dimensions. *Int. J. Bif. Chaos*, 1:493–520, 1991a.

E. Doedel, H. B. Keller, and J. P. Kernevez. Numerical analysis and control of bifurcation problems (II) Bifurcation in infinite dimensions. *Int. J. Bif. Chaos*, 1:745–72, 1991b.

E. J. Doedel, A. R. Champneys, T. F. Fairgrieve, Y. A. Kuznetsov, B. Sandstede, and X. J. Wang. *AUTO97: Continuation and bifurcation software for ordinary differential equations*. Concordia University, Montreal, 1997.

F. Domingues Dos Santos and T. Ondarçuhu. Free-running droplets. *Phys. Rev. Lett.*, 75:2972–2975, 1995.

B. Y. Du, F. C. Xie, Y. J. Wang, Z. Y. Yang, and O. K. C. Tsui. Dewetting of polymer films with built-in topographical defects. *Langmuir*, 18:8510–8517, 2002.

E. B. Dussan. On the spreading of liquids on solid surfaces: Static and dynamic contact lines. *Ann. Rev. Fluid Mech.*, 11:371–400, 1979.

I. E. Dzyaloshinskii, E. M. Lifshitz, and L. P. Pitaevskii. Van der Waals forces in liquid films. *Sov. Phys. JETP*, 37:161, 1960.

J. Eggers. Nonlinear dynamics and breakup of free-surface flows. *Rev. Mod. Phys.*, 69: 865–929, 1997.

J. Eggers. Hydrodynamic theory of forced dewetting. *Phys. Rev. Lett.*, 93:094502, 2004.

M. Elbaum and S. G. Lipson. How does a thin wetted film dry up? *Phys. Rev. Lett.*, 72:3562–3565, 1994.

M. H. Eres, L. W. Schwartz, and R. V. Roy. Fingering phenomena for driven coating films. *Phys. Fluids*, 12:1278–1295, 2000.

A. Faldi, R. J. Composto, and K. I. Winey. Unstable polymer bilayers. 1. Morphology of dewetting. *Langmuir*, 11:4855, 1995.

H. Gau, S. Herminghaus, P. Lenz, and R. Lipowsky. Liquid morphologies on structured surfaces: From microchannels to microchips. *Science*, 283:46–49, 1999.

D. P. III Gaver and J. B. Grotberg. The dynamics of a localized surfactant on a thin film. *J. Fluid Mech.*, 213:127–148, 1990.

B. Gjevik. Occurrence of finite-amplitude surface waves on falling liquid films. *Phys. Fluids*, 13:1918–1925, 1970.

K. B. Glaser and T. P. Witelski. Coarsening dynamics of dewetting films. *Phys. Rev. E*, 67:016302, 2003.

A. A. Golovin, A. A. Nepomnyashchy, S. H. Davis, and M. A. Zaks. Convective Cahn-Hilliard models: From coarsening to roughening. *Phys. Rev. Lett.*, 86:1550–1553, 2001.

A. A. Golovin, A. A. Nepomnyashchy, and L. M. Pismen. Interaction between short-scale Marangoni convection and long-scale deformational instability. *Phys. Fluids*, 6:34–48, 1994.

H. P. Greenspan. On the motion of a small viscous droplet that wets a surface (relevant to cell movement). *J. Fluid Mech.*, 84:125–143, 1978.

G. Grün, K. Mecke, and M. Rauscher. Thin film flow influenced by thermal noise. *preprint*, 2005. submitted.

J. Guckenheimer and P. Holmes. *Nonlinear Oscillations, Dynamical Systems and Bifurkations of Vector Fields*, volume 42 of *Applied Mathematical Sciences*. Springer-Verlag, Berlin, 1993.

W. B. Hardy. Historical notes upon surface energy and forces of short range. *Nature*, 109:375–378, 1922.

F. Hauksbee. Several experiments touching the seeming spontaneous ascent of water. *Phil. Trans.*, 26:258–266, 1708.

F. Hauksbee. An account of an experiment touching the direction of a drop of oil of oranges, between two glass planes, towards any side of them that is nearest press'd together. *Phil. Trans.*, 27:395–396, 1710.

Y. Hayashima, M. Nagayama, Y. Doi, S. Nakata, M. Kimura, and M. Iida. Self-motion of a camphoric acid boat sensitive to the chemical environment. *Phys. Chem. Chem. Phys.*, 4:1386–1392, 2002.

S. Herminghaus, A. Fery, S. Schlagowski, K. Jacobs, R. Seemann, H. Gau, W. Mönch, and T. Pompe. Liquid microstructures at solid interfaces. *J. Phys.-Condes. Matter*, 12:A57–A74, 2000.

L. M. Hocking. A moving fluid interface. II. The removal of the force singularity by a slip flow. *J. Fluid Mech.*, 79:209–229, 1977.

L. M. Hocking. Spreading and instability of a viscous fluid sheet. *J. Fluid Mech.*, 211: 373–392, 1990.

L. M. Hocking and M. J. Miksis. Stability of a ridge of fluid. *J. Fluid Mech.*, 247:157–177, 1993.

C. Huh and L. E. Scriven. Hydrodynamic model of steady movement of a solid / liquid / fluid contact line. *J. Colloid Interface Sci.*, 35:85–101, 1971.

R. J. Hunter. *Foundation of Colloid Science*, volume 1. Clarendon Press, Oxford, 1992.

H. E. Huppert. Flow and instability of a viscous current down a slope. *Nature*, 300: 427–429, 1982.

J. N. Israelachvili. *Intermolecular and Surface Forces*. Academic Press, London, 1992.

K. Jacobs, S. Herminghaus, and K. R. Mecke. Thin liquid polymer films rupture via defects. *Langmuir*, 14:965–969, 1998.

K. John, M. Bär, and U. Thiele. Self-propelled running droplets on solid substrates driven by chemical reactions. *Eur. Phys. J. E*, 18:183–199, 2005.

S. W. Joo, S. H. Davis, and S. G. Bankoff. Long-wave instabilities of heated falling films: Two-dimensional theory of uniform layers. *J. Fluid Mech.*, 230:117–146, 1991.

O. A. Kabov and I. V. Marchuk. Infrared study of the liquid film flowing on surface with nonuniform heat flux distribution. *Heat Transfer Research*, 29:544–562, 1998.

S. Kalliadasis. Nonlinear instability of a contact line driven by gravity. *J. Fluid Mech.*, 413:355–378, 2000.

P. L. Kapitza. Waveflow of thin layers of a viscous fluid: I. The free flow. *Zh. Exp. Teor. Fiz.*, 18:3–18, 1949.

P. L. Kapitza and S. P. Kapitza. Waveflow of thin layers of a viscous fluid: III. Experimental study of undulatory flow conditions. *Zh. Exp. Teor. Fiz.*, 19:105–120, 1949.

K. Kargupta, R. Konnur, and A. Sharma. Instability and pattern formation in thin liquid films on chemically heterogeneous substrates. *Langmuir*, 16:10243–10253, 2000.

K. Kargupta, R. Konnur, and A. Sharma. Spontaneous dewetting and ordered patterns in evaporating thin liquid films on homogeneous and heterogeneous substrates. *Langmuir*, 17:1294–1305, 2001.

K. Kargupta and A. Sharma. Templating of thin films induced by dewetting on patterned surfaces. *Phys. Rev. Lett.*, 86:4536–4539, 2001.

K. Kargupta and A. Sharma. Creation of ordered patterns by dewetting of thin films on homogeneous and heterogeneous substrates. *J. Colloid Interface Sci.*, 245:99–115, 2002.

A. Karim, J. F. Douglas, B. P. Lee, S. C. Glotzer, J. A. Rogers, R. J. Jackman, E. J. Amis, and G. M. Whitesides. Phase separation of ultrathin polymer-blend films on patterned substrates. *Phys. Rev. E*, 57:R6273–R6276, 1998.

D. E. Kataoka and S. M. Troian. A theoretical study of instabilities at the advancing front of thermally driven coating films. *J. Colloid Interface Sci.*, 192:350–362, 1997.

D. E. Kataoka and S. M. Troian. Stabilizing the advancing front of thermally driven climbing films. *J. Colloid Interface Sci.*, 203:335–344, 1998.

I. G. Kevrekidis, B. Nicolaenko, and J. C. Scovel. Back in the saddle again - a computer-assisted study of the Kuramoto-Sivashinsky equation. *SIAM J. Appl. Math.*, 50:760–790, 1990.

H. S. Kheshgi and L. E. Scriven. Dewetting: Nucleation and growth of dry regions. *Chem. Eng. Sci.*, 46:519–526, 1991.

H. I. Kim, C. M. Mate, K. A. Hannibal, and S. S. Perry. How disjoining pressure drives the dewetting of a polymer film on a silicon surface. *Phys. Rev. Lett.*, 82:3496–3499, 1999.

A. Knoll, A. Horvat, K. S. Lyakhova, G. Krausch, G. J. A. Sevink, A. V. Zvelindovsky, and R. Magerle. Phase behavior in thin films of cylinder-forming block copolymers. *Phys. Rev. Lett.*, 89:035501, 2002.

R. Konnur, K. Kargupta, and A. Sharma. Instability and morphology of thin liquid films on chemically heterogeneous substrates. *Phys. Rev. Lett.*, 84:931–934, 2000.

S. Krishnamoorthy, B. Ramaswamy, and S. W. Joo. Spontaneous rupture of thin liquid films due to thermocapillarity: A full-scale direct numerical simulation. *Phys. Fluids*, 7:2291–2293, 1995.

Y. Kuramoto and T. Tsuzuki. Persistent propagation of concentration waves in dissipative media far from thermal equilibrium. *Prog. Theor. Phys.*, 55:356–369, 1976.

P. Lambooy, K. C. Phelan, O. Haugg, and G. Krausch. Dewetting at the liquid-liquid interface. *Phys. Rev. Lett.*, 76:1110–1113, 1996.

J. S. Langer. *An introduction to the kinetics of first-order phase transitions*, chapter 3, pages 297–363. Cambridge University Press, 1992.

P. S. Laplace. Sur l'action capillaire. *Suppl. au livre X, Traité de Mécanique Céleste*, page 349, 1806.

S. W. Lee, D. Y. Kwok, and P. E. Laibinis. Chemical influences on adsorption-mediated self-propelled drop movement. *Phys. Rev. E*, 65:051602, 2002.

P. Lenz and R. Lipowsky. Morphological transitions of wetting layers on structured surfaces. *Phys. Rev. Lett.*, 80:1920–1923, 1998.

S. P. Lin. Finite amplitude side-band instability of a viscous film. *J. Fluid Mech.*, 63: 417–429, 1974.

S. P. Lin and H. Brenner. Tear film rupture. *J. Colloid Interface Sci.*, 89:226–231, 1982.

J. Liu and J. P. Gollub. Solitary wave dynamics of film flows. *Phys. Fluids*, 6:1702–1712, 1994.

C. G. Marangoni. Ueber die Ausbreitung der Tropfen einer Flüssigkeit auf der Oberfläche einer anderen. *Ann. Phys. (Poggendorf)*, 143:337–354, 1871.

J. L. Masson, O. Olufokunbi, and P. F. Green. Flow instabilities in entangled polymer thin films. *Macromolecules*, 35:6992–6996, 2002.

O. K. Matar, R. V. Craster, and M. R. E. Warner. Surfactant transport on highly viscous surface films. *J. Fluid Mech.*, 466:85–111, 2002.

O. K. Matar and S. M. Troian. Linear stability analysis of an insoluble surfactant monolayer spreading on a thin liquid film. *Phys. Fluids*, 9:3645–3657, 1997.

J. C. Meredith, A. P. Smith, A. Karim, and E. J. Amis. Combinatorial materials science for polymer thin-film dewetting. *Macromolecules*, 33:9747–9756, 2000.

D. Merkt, A. Pototsky, M. Bestehorn, and U. Thiele. Long-wave theory of bounded two-layer films with a free liquid-liquid interface: Short- and long-time evolution. *Phys. Fluids*, 17:064104, 2005.

E. C. Millington. Studies in capillarity and cohesion in the eighteenth century. *Annals of Science*, 5:352–369, 1945.

V. S. Mitlin. Dewetting of solid surface: Analogy with spinodal decomposition. *J. Colloid Interface Sci.*, 156:491–497, 1993.

V. S. Mitlin. Dewetting revisited: New asymptotics of the film stability diagram and the metastable regime of nucleation and growth of dry zones. *J. Colloid Interface Sci.*, 227:371–379, 2000.

V. S. Mitlin. Numerical study of Lifshitz-Slyozov-like metastable dewetting model. *J. Colloid Interface Sci.*, 233:153–158, 2001.

M. D. Morariu, E. Schäffer, and U. Steiner. Capillary instabilities by fluctuation induced forces. *Eur. Phys. J. E*, 12:375–379, 2003.

D. T. Moyle, M.-S. Chen, and G. M. Homsy. Nonlinear rivulet dynamics during unstable wetting flows. *Int. J. Multiphase Flow*, 25:1243–1262, 1999.

A. Münch. Dewetting rates of thin liquid films. *J. Phys.-Condes. Matter*, 17:S309–S318, 2005.

A. A. Nepomnyashchy, M. G. Velarde, and P. Colinet. *Interfacial Phenomena and Convection.* Chapman & Hall/CRC, Boca Raton, 2002.

I. Newton. *Opticks.* G. Bell & Sons LTD., London, 1730a. (reprinted 4th ed. 1931, Book II, Part 1, Obs. 17-19).

I. Newton. *Opticks.* G. Bell & Sons LTD., London, 1730b. (reprinted 4th ed. 1931, Book III, Part 1, Querie 31).

L. T. Nguyen and V. Balakotaiah. Modeling and experimental studies of wave evolution on free falling viscous films. *Phys. Fluids*, 12:2236–2256, 2000.

A. Novick-Cohen. The nonlinear Cahn - Hilliard equation: Transition from spinodal decomposition to nucleation behavior. *J. Stat. Phys.*, 38:707–723, 1985.

A. Oron. Three-dimensional nonlinear dynamics of thin liquid films. *Phys. Rev. Lett.*, 85:2108–2111, 2000.

A. Oron, S. H. Davis, and S. G. Bankoff. Long-scale evolution of thin liquid films. *Rev. Mod. Phys.*, 69:931–980, 1997.

A. Oron and P. Rosenau. Formation of patterns induced by thermocapillarity and gravity. *J. Physique II France*, 2:131–146, 1992.

Q. Pan, K. I. Winey, H. H. Hu, and R. J. Composto. Unstable polymer bilayers. 2. The effect of film thickness. *Langmuir*, 13:1758–1766, 1997.

V. N. Paunov, K. D. Danov, N. Alleborn, H. Raszillier, and F. Durst. Stability of evaporating two-layered liquid film in the presence of surfactant - iii. non-linear stability analysis. *Chem. Eng. Sci.*, 53:2839–2857, 1998.

L. M. Pismen. Nonlocal diffuse interface theory of thin films and the moving contact line. *Phys. Rev. E*, 6402:021603, 2001.

L. M. Pismen and Y. Pomeau. Disjoining potential and spreading of thin liquid layers in the diffuse interface model coupled to hydrodynamics. *Phys. Rev. E*, 62:2480–2492, 2000.

L. M. Pismen and Y. Pomeau. Mobility and interactions of weakly nonwetting droplets. *Phys. Fluids*, 16:2604–2612, 2004.

L. M. Pismen and U. Thiele. Asymptotic theory for a moving droplet driven by a wettability gradient. *Phys. Fluids*, 2006. (in press).

J. A. F. Plateau. *Statique Expérimentale et Théorique des Liquides Soumis aux Seules Forces Moléculaires*. Gauthier-Villars, Paris, 1873.

I. Podariu, Z. Y. Shou, and A. Chakrabarti. Viscous flow and coarsening of microdomains in diblock copolymer thin films. *Phys. Rev. E*, 62:R3059–R3062, 2000.

T. Podgorski. Ruisselement en condition de mouillage partiel, 2000. PhD Thesis (U. Paris 6).

T. Podgorski, J.-M. Flesselles, and L. Limat. Corners, cusps, and pearls in running drops. *Phys. Rev. Lett.*, 87:036102, 2001.

Y. Pomeau. Représentation de la ligne de contact mobile dans les équations de la mécanique des fluides. *C. R. Acad. Sci. Ser. II-B*, 328:411–416, 2000.

A. Pototsky, M. Bestehorn, D. Merkt, and U. Thiele. Alternative pathways of dewetting for a thin liquid two-layer film. *Phys. Rev. E*, 70:025201(R), 2004.

A. Pototsky, M. Bestehorn, D. Merkt, and U. Thiele. Morphology changes in the evolution of liquid two-layer films. *J. Chem. Phys.*, 122:224711, 2005.

R. F. Probstein. *Physicochemical Hydrodynamics*. Wiley, New York, 2. edition, 1994.

A. Pumir, P. Manneville, and Y. Pomeau. On solitary waves running down an inclined plane. *J. Fluid Mech.*, 135:27–50, 1983.

D. Quéré, M. J. Azzopardi, and L. Delattre. Drops at rest on a tilted plane. *Langmuir*, 14:2213–2216, 1998.

J. W. S. Rayleigh. On convective currents in a horizontal layer of fluid, when the higher temperature is on the under side. *Phil. Mag. S.6*, 32:529–546, 1916.

C. Redon, F. Brochard-Wyart, and F. Rondelez. Dynamics of dewetting. *Phys. Rev. Lett.*, 66:715–718, 1991.

N. Rehse, C. Wang, M. Hund, M. Geoghegan, R. Magerle, and G. Krausch. Stability of thin polymer films on a corrugated substrate. *Eur. Phys. J. E*, 4:69–76, 2001.

G. Reiter. Dewetting of thin polymer films. *Phys. Rev. Lett.*, 68:75–78, 1992.

G. Reiter. Unstable thin polymer films: Rupture and dwetting. *Langmuir*, 9:1344, 1993.

G. Reiter. Dewetting of highly elastic thin polymer films. *Phys. Rev. Lett.*, 87:186101, 2001.

G. Reiter and A. Sharma. Auto-optimization of dewetting rates by rim instabilities in slipping polymer films. *Phys. Rev. Lett.*, 8716:166103, 2001.

C. Renger, P. Müller-Buschbaum, M. Stamm, and G. Hinrichsen. Investigation and retardation of the dewetting on top of highly viscous amorphous substrates. *Macromolecules*, 33:8388–8398, 2000.

O. Reynolds. On the theory of lubrication and its application to Mr. Beauchamp Tower's experiments, including an experimental determination of the viscosity of olive oil. *Phil. Trans. Roy. Soc.*, 177:157–234, 1886.

L. Rockford, Y. Liu, P. Mansky, T. P. Russell, M. Yoon, and S. G. J. Mochrie. Polymers on nanoperiodic, heterogeneous surfaces. *Phys. Rev. Lett.*, 82:2602–2605, 1999.

R. V. Roy and L. W. Schwartz. On the stability of liquid ridges. *J. Fluid Mech.*, 391: 293–318, 1999.

E. Ruckenstein and R. K. Jain. Spontaneous rupture of thin liquid films. *J. Chem. Soc. Faraday Trans. II*, 70:132–147, 1974.

T. R. Salamon, R. C. Armstrong, and R. A. Brown. Traveling waves on vertical films: Numerical analysis using the finite element method. *Phys. Fluids*, 5:2202–2220, 1994.

N. Samid-Merzel, S. G. Lipson, and D. S. Tannhauser. Pattern formation in drying water films. *Phys. Rev. E*, 57:2906–2913, 1998.

E. Schäffer, S. Harkema, M. Roerdink, R. Blossey, and U. Steiner. Morphological instability of a confined polymer film in a thermal gradient. *Macromolecules*, 36:1645–1655, 2003.

E. Schäffer and U. Steiner. Acoustic instabilities in thin polymer films. *Eur. Phys. J. E*, 8:347–351, 2002.

B. Scheid, A. Oron, P. Colinet, U. Thiele, and J. C. Legros. Nonlinear evolution of nonuniformly heated falling liquid films. *Phys. Fluids*, 14:4130–4151, 2002.

B. Scheid, C. Ruyer-Quil, U. Thiele, O. A. Kabov, J. C. Legros, and P. Colinet. Validity domain of the Benney equation including Marangoni effect for closed and open flows. *J. Fluid Mech.*, 527:303–335, 2005.

S. Schlagowski, K. Jacobs, and S. Herminghaus. Nucleation-induced undulative instability in thin films of nCB liquid crystals. *Europhys. Lett.*, 57:519–525, 2002.

A. K. Schmid, N. C. Bartelt, and R. Q. Hwang. Alloying at surfaces by the migration of reactive twodimensional islands. *Science*, 290:1561–1564, 2000.

L. E. Scriven and C. V. Sternling. Marangoni effects. *Nature*, 187:186–188, 1960.

R. Seemann, S. Herminghaus, and K. Jacobs. Dewetting patterns and molecular forces: A reconciliation. *Phys. Rev. Lett.*, 86:5534–5537, 2001a.

R. Seemann, S. Herminghaus, and K. Jacobs. Shape of a liquid front upon dewetting. *Phys. Rev. Lett.*, 87:196101, 2001b.

R. Seemann, S. Herminghaus, C. Neto, S. Schlagowski, D. Podzimek, R. Konrad, H. Mantz, and K. Jacobs. Dynamics and structure formation in thin polymer melt films. *J. Phys.-Condes. Matter*, 17:S267–S290, 2005.

A. Sehgal, V. Ferreiro, J. F. Douglas, E. J. Amis, and A. Karim. Pattern-directed dewetting of ultrathin polymer films. *Langmuir*, 18:7041–7048, 2002.

K. Sekimoto, R. Oguma, and K. Kawasaki. Morphological stability analysis of partial wetting. *Ann. Phys.*, 176:359–392, 1987.

M. Sferrazza, M. Heppenstall-Butler, R. Cubitt, D. Bucknall, J. Webster, and R. A. L. Jones. Interfacial instability driven by dispersive forces: The early stages of spinodal dewetting of a thin polymer film on a polymer substrate. *Phys. Rev. Lett.*, 81:5173–5176, 1998.

M. Sferrazza, C. Xiao, R. A. L. Jones, D. G. Bucknall, J. Webster, and J. Penfold. Evidence for capillary waves at immiscible polymer/polymer interfaces. *Phys. Rev. Lett.*, 78:3693–3696, 1997.

A. Sharma. Equilibrium contact angles and film thicknesses in the apolar and polar systems: Role of intermolecular interactions in coexistence of drops with thin films. *Langmuir*, 9:3580, 1993a.

A. Sharma. Relationship of thin film stability and morphology to macroscopic parameters of wetting in the apolar and polar systems. *Langmuir*, 9:861–869, 1993b.

A. Sharma and A. T. Jameel. Nonlinear stability, rupture and morphological phase separation of thin fluid films on apolar and polar substrates. *J. Colloid Interface Sci.*, 161:190–208, 1993.

A. Sharma and J. Mittal. Instability of thin liquid films by density variations: A new mechanism that mimics spinodal dewetting. *Phys. Rev. Lett.*, 89:186101, 2002.

A. Sharma and G. Reiter. Instability of thin polymer films on coated substrates: Rupture, dewetting and drop formation. *J. Colloid Interface Sci.*, 178:383–399, 1996.

A. Sharma and E. Ruckenstein. Mechanism of tear film rupture and its implications for contact-lens tolerance. *Amer. J. Optom. Physiol. Opt.*, 62:246–253, 1985.

W. Ya. Shkadov. Wave conditions in the flow of a thin layer of a viscous liquid under the action of gravity. *Izv. Akad. Nauk SSSR, Mekh. Zhidk. Gaza*, 1:43–51, 1967.

N. Silvi and V. E. B. Dussan. The rewetting of an inclined solid surface by a liquid. *Phys. Fluids*, 28:5–7, 1985.

G. I. Sivashinsky. Non-linear analysis of hydrodynamic instability in laminar flames. 1. Derivation of basic equations. *Acta Astronaut.*, 4:1177–1206, 1977.

J. M. Skotheim, U. Thiele, and B. Scheid. On the instability of a falling film due to localized heating. *J. Fluid Mech.*, 475:1–19, 2003.

A. Sommerfeld. Zur hydrodynamischen Theorie der Schmiermittelreibung. *Z. Math. Phys.*, 50:97–155, 1904.

M. A. Spaid and G. M. Homsy. Stability of Newtonian and viscoelastic dynamic contact lines. *Phys. Fluids*, 8:460–478, 1996.

Y. Sumino, N. Magome, T. Hamada, and K. Yoshikawa. Self-running droplet: Emergence of regular motion from nonequilibrium noise. *Phys. Rev. Lett.*, 94(6):068301, 2005a.

Y. Sumino, M. Nagayama, H. Kitahata, S.-i.M. Nomura, N. Magome, Y. Mori, and K. Yoshikawa. Chemo-sensitive running droplet. *arXiv:nlin.AO/0505006*, 2005b.

M. J. Tan, S. G. Bankoff, and S. H. Davis. Steady thermocapillary flows of thin liquid layers. I. Theory. *Phys. Fluids A*, 2:313–321, 1990.

G. F. Teletzke, H. T. Davis, and L. E. Scriven. Wetting hydrodynamics. *Rev. Phys. Appl.*, 23:989–1007, 1988.

U. Thiele. Entnetzung von Kollagenfilmen, 1998. PhD-thesis, Dresden.

U. Thiele. Open questions and promising new fields in dewetting. *Eur. Phys. J. E*, 12: 409–416, 2003a.

U. Thiele. Tentative interpretation of the dewetting morphologies presented by Tsui et al. *Eur. Phys. J. E*, 12:427–430, 2003b.

U. Thiele, L. Brusch, M. Bestehorn, and M. Bär. Modelling thin-film dewetting on structured substrates and templates: Bifurcation analysis and numerical simulations. *Eur. Phys. J. E*, 11:255–271, 2003.

U. Thiele, K. John, and M. Bär. Dynamical model for chemically driven running droplets. *Phys. Rev. Lett.*, 93:027802, 2004.

U. Thiele and E. Knobloch. Front and back instability of a liquid film on a slightly inclined plate. *Phys. Fluids*, 15:892–907, 2003.

U. Thiele and E. Knobloch. Thin liquid films on a slightly inclined heated plate. *Physica D*, 190:213–248, 2004.

U. Thiele, M. Mertig, and W. Pompe. Dewetting of an evaporating thin liquid film: Heterogeneous nucleation and surface instability. *Phys. Rev. Lett.*, 80:2869–2872, 1998.

U. Thiele, K. Neuffer, M. Bestehorn, Y. Pomeau, and M. G. Velarde. Sliding drops on an inclined plane. *Colloid Surf. A*, 206:87–104, 2002a.

U. Thiele, K. Neuffer, Y. Pomeau, and M. G. Velarde. On the importance of nucleation solutions for the rupture of thin liquid films. *Colloid Surf. A*, 206:135–155, 2002b.

U. Thiele, M. G. Velarde, and K. Neuffer. Dewetting: Film rupture by nucleation in the spinodal regime. *Phys. Rev. Lett.*, 87:016104, 2001a.

U. Thiele, M. G. Velarde, K. Neuffer, M. Bestehorn, and Y. Pomeau. Sliding drops in the diffuse interface model coupled to hydrodynamics. *Phys. Rev. E*, 64:061601, 2001b.

U. Thiele, M. G. Velarde, K. Neuffer, and Y. Pomeau. Film rupture in the diffuse interface model coupled to hydrodynamics. *Phys. Rev. E*, 64:031602, 2001c.

J. Thomson. On certain curious motions observable at the surface of wine and other alcoholic liquors. *Phil. Mag. Ser. 4*, 10:330–333, 1855.

W. Thomson. On the division of space with minimum partitional area. *Acta Math.*, 11: 121–134, 1887.

C. Tomlinson. On the motion of certain liquids on the surface of water. *Phil. Mag. Ser. 4*, 39:32–48, 1870.

S. M. Troian, E. Herbolzheimer, S. A. Safran, and J. F. Joanny. Fingering instabilities of driven spreading films. *Europhys. Lett.*, 10:25–30, 1989.

M. W. J. van der Wielen, E. P. I. Baars, M. Giesbers, M. A. C. Stuart, and G. J. Fleer. The effect of substrate modification on the ordering and dewetting behavior of thin liquid-crystalline polymer films. *Langmuir*, 16:10137–10143, 2000.

S. J. VanHook, M. F. Schatz, W. D. McCormick, J. B. Swift, and H. L. Swinney. Long-wavelength instability in surface-tension-driven Bénard convection. *Phys. Rev. Lett.*, 75:4397–4400, 1995.

S. J. VanHook, M. F. Schatz, J. B. Swift, W. D. McCormick, and H. L. Swinney. Long-wavelength surface-tension-driven Bénard convection: Experiment and theory. *J. Fluid Mech.*, 345:45–78, 1997.

I. Veretennikov, A. Indeikina, and H.-C. Chang. Front dynamics and fingering of a driven contact line. *J. Fluid Mech.*, 373:81–110, 1998.

A. Vrij. Possible mechanism for the spontaneous rupture of thin free liquid films. *Disc. Faraday Soc.*, 42:23–33, 1966.

R. Xie, A. Karim, J. F. Douglas, C. C. Han, and R. A. Weiss. Spinodal dewetting of thin polymer films. *Phys. Rev. Lett.*, 81:1251–1254, 1998.

Y. Ye and H.-C. Chang. A spectral theory for fingering on a prewetted plane. *Phys. Fluids*, 11:2494–2515, 1999.

R. Yerushalmi-Rozen, T. Kerle, and J. Klein. Alternative dewetting pathways of thin liquid films. *Science*, 285:1254–1256, 1999.

C. S. Yih. Stability of liquid flow down an inclined plane. *Phys. Fluids*, 6:321–334, 1963.

T. Young. An essay on the cohesion of fluids. *Phil. Trans. R. Soc.*, 95:65–87, 1805.

Y. L. Zhang, O. K. Matar, and R. V. Craster. Analysis of tear film rupture: effect of non-newtonian rheology. *J. Colloid Interface Sci.*, 262:130–148, 2003.

D. W. Zheng, W. Wen, and K. N. Tu. Reactive wetting- and dewetting-induced diffusion-limited aggregation. *Phys. Rev. E*, 57:R3719–R3722, 1998.

P. Ziherl, R. Podgornik, and S. Zumer. Pseudo-casimir structural force drives spinodal dewetting in nematic liquid crystals. *Phys. Rev. Lett.*, 84:1228–1231, 2000.

Singularities and Similarities

Jens G. Eggers

School of Mathematics, University of Bristol, Bristol, United Kingdom

Abstract We discuss the mathematical description of self-similar phenomena. Such phenomena cover many different length or time scales. Self-similarity allows a transformation from one scale to another and final reduction to a system that has only one scale. This is illustrated through a series of examples including diffusion processes, drop pinch-off and nanojets.

1 Introduction

We focus on phenomena which cover many different length or time scales. Namely, it is extremely common that parts of the solution 'living' on different scales look identical under a rescaling of their dimensions: they are self-similar. Some of the reasoning that explains this observation is given below, but the ubiquity of such scaling phenomena remains remarkable. More pragmatically, self-similarity is the reason why many-scaled phenomena are amenable to a theoretical description in the first place. Simply put, it means they can be reduced to something that has only one scale, and the complexity is all in the transformation from one scale to the other.

The examples and illustrations given below are all taken from hydrodynamics; other examples (without much danger of overlap) could have been taken from completely different fields of mathematics, physics, or engineering. In particular, I focus on singularities, i.e. the development of arbitrarily small structures in finite time. One such example occurs when a fluid drop falls from a faucet: at some point the thickness of the fluid neck separating the drop from the faucet must go to zero. As the singularity is approached, smaller and smaller length scales are covered, until the scale of molecules is reached.

Many of the points and observations made below are discussed in much more detail in the beautiful book by Barenblatt (1996). I also describe an instance where strict self-similarity breaks down, and the concept needs to be generalized. Such generalizations are an important topic of current research. In the end I discuss drop breakup on a very small scale, such that thermal fluctuations become important: self-similarity can then only be valid in a suitably defined averaged sense.

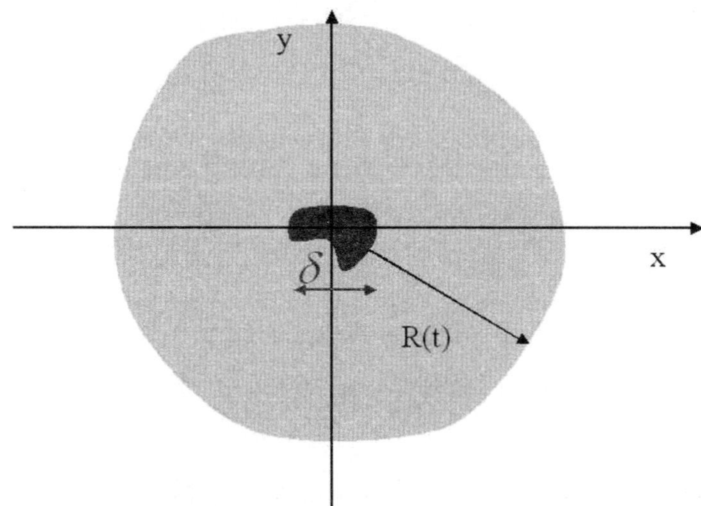

Figure 1. A spreading drop whose initial size was δ and of arbitrary shape.

2 A Simple Example: The Diffusion Equation

An initial drop of contaminant (typical size δ) spreads diffusively in a container of size L. The spreading law is:

$$\frac{\partial u}{\partial t} = \kappa \left(\frac{\partial^2 u}{\partial x^2} + \frac{\partial^2 u}{\partial y^2} \right) \equiv \kappa \Delta u. \tag{2.1}$$

What does the solution for the contaminant density $u(x,t)$ look like? Plausibly the long–term solution can depend only on the amount of 'stuff' Q, since this quantity is *conserved* during the evolution. From the initial value one obtains

$$Q = \int_{\mathbb{R}^2} u(x,0)d^2x.$$

Letting the dimension of Q be mass, the dimensions of the other quantities become:

$$[u] = \frac{g}{cm^2}, \quad [\kappa] = \frac{cm^2}{s}, \quad [x] = [y] = [\delta] = [L] = cm.$$

The most general form of the spreading law is

$$u = f(\underbrace{Q, \kappa, t, x, y, \delta, L}..),$$

$$7 \text{ arguments}$$

a mess!

First we use *dimensional analysis*, making everything *dimensionless* using characteristic length and mass scales in the problem. Namely: $\ell = \sqrt{\kappa t}$ (involving the diffusion constant) and $u_0 = Q/\ell^2$:

$$u = u_0 F\left(\frac{x}{\ell}, \frac{y}{\ell}, \frac{\delta}{\ell}, \frac{L}{\ell}\right).$$

Since this result is invariant under the choice of 3 arbitrary dimensional scales (cm, s, g) the number of arguments is now $4 = 7 - 3$.

Second we use physical reasoning to further reduce the number of parameters. If the drop has spread to a much greater size than δ, yet is still much smaller than the size of the box L, $\delta/\ell \ll 1, L/\ell \gg 1$, then the dynamics should have 'forgotten' about the initial condition, but doesn't yet 'feel' the outer boundary: F does not depend on δ/ℓ or L/ℓ. More formally: we replace the arguments by their limits and hope that F remains finite!

$$u = u_0 F\left(\frac{x}{\ell}, \frac{y}{\ell}, 0, \infty\right) \equiv u_0 \bar{F}\left(\frac{x}{\ell}, \frac{y}{\ell}\right).$$

We call

$$u = \frac{Q}{\kappa t} \bar{F}\left(\frac{x}{\sqrt{\kappa t}}, \frac{y}{\sqrt{\kappa t}}\right) \tag{2.2}$$

a similarity solution. A more general structure is

$$u = f(t) F\left(\frac{x_1}{f_1(t)}, \frac{x_2}{f_2(t)}, \cdots\right)$$

because a change in t only results in the *rescaling* of dependent and independent variables. In most cases $f_i(t)$ is a power law; below we are going to say a few words as to why power laws are special.

Now if u has 'forgotten' about the initial condition $u(\boldsymbol{x},0)$, $\bar{F}$ must depend on x/ℓ and y/ℓ in the same way: any anisotropy can only come in through the initial condition. Thus

$$u = \frac{Q}{\kappa t} G\left(\frac{|\boldsymbol{x}|}{\sqrt{\kappa t}}\right), \quad \xi := \frac{|\boldsymbol{x}|}{\sqrt{\kappa t}} \tag{2.3}$$

is the form of the desired similarity solution.

Plugging this into the original equation, we find an equation for G:

$$-\frac{Q}{\kappa t^2} G - \frac{Q}{2\kappa t} \frac{|x|}{\sqrt{\kappa}t^{3/2}} G' = \frac{\kappa Q}{\kappa^2 t^2} \frac{1}{\xi}(\xi G')', \qquad \left[\text{note:} \quad \Delta f(r) = \frac{1}{r}(rf')'\right]$$

hence

$$-G' - \frac{\xi}{2} G' = \frac{1}{\xi}(\xi G'), \tag{2.4}$$

which is a called a *similarity equation.*

Equation (2.4) is easily solved:

$$-\frac{1}{2}(\xi^2 G')' = (\xi G')' \rightarrow -\frac{\xi^2}{2} G = \xi G' + C, \text{ put } \xi = 0 \rightarrow C = 0$$

$$G' = -\frac{\xi G}{2} \rightarrow G = G_0 e^{-\xi^2/4},$$

and thus

$$u_{sim}(\boldsymbol{x}, t) = \frac{QG_0}{\kappa t} e^{-\frac{|\boldsymbol{x}|^2}{4\kappa t}},$$ (2.5)

which is the explicit form of the similarity solution we have been seeking. The total mass Q is given by

$$\int_{\mathbb{R}^2} u d^2 x = 4\pi Q G_0,$$

and thus $G_0 = 1/4\pi$.

The typical radius of the drop is

$$R = \frac{1}{Q} \int_{\mathbb{R}^2} |\mathbf{x}| u_{sim}(|\mathbf{x}|, t) d^2 x = \frac{1}{2\kappa t} \int_0^\infty r^2 e^{-\frac{r^2}{4\kappa t}} dr =$$

$$\frac{(4\kappa t)^{3/2}}{(2\kappa t)} \underbrace{\int_0^\infty \xi^2 e^{-\xi^2} d\xi}_{\sqrt{\pi}/4} = \sqrt{\pi\kappa} \;\; t^{1/2},$$ (2.6)

and thus behaves like a power law. According to our reasoning, the power law should be valid for $\delta \ll R \ll L$.

<u>Excursion:</u> What is special about power laws?

Consider

$$R = Mt^\alpha.$$ (2.7)

The main point is that power laws are scale-invariant, and thus do not specify a particular scale. Any change of t-scale (t_0) can be absorbed into a change of R-scale (R_0):

$$R = Mt_0^\alpha \, (t/t_0)^\alpha, \; R_0 \equiv t_0^\alpha$$

and thus

$$\tilde{R} = M\tilde{t}^\alpha,$$

which is the same as (2.7), but with $\tilde{R} = R/R_0$ and $\tilde{t} = t/t_0$.

This property also <u>characterizes</u> power laws: let

$$R = f(t) \quad \text{and} \quad R/R_0(t_0) = f(t/t_0),$$

which is the general condition for scale invariance. It follows that

$$R(t) = f(1)R_0(t), \; \text{thus } R(t)/R(t_0) = f(t/t_0)/f(1).$$

Now differentiate with respect to t: $R'(t)/R(t_0) = f'/(t_0 f(1))$, and put $t = t_0$. This gives

$$(\ln R)' = \frac{f'(1)}{f(1)} (\ln t)'$$

and thus $R = R_0 t^\alpha$, where $\alpha = f'(1)/f(1)$ and R_0 is a constant of integration.

Now two more remarks about the results obtained so far are in order:
Remark 1 A crucial ingredient in finding the exponents in the scaling ansatz

$$u = t^\alpha G\left(\frac{r}{t^\beta}\right) \tag{2.8}$$

(namely $\alpha = -1, \beta = 1/2$, cf. equation 2.3) was the assumption of *regularity* in the limits $\delta/\ell \to 0$ and $L/\ell \to \infty$. It resulted in rational exponents determined by dimensional analysis alone. If $F(r/\ell, \delta/\ell)$ had not been regular in the limit $\delta/\ell \to 0$, while keeping with the assumption of self-similarity, the answer would have been:

$$u = u_0 F\left(\frac{r}{\ell}\left(\frac{\delta}{\ell}\right)^\gamma\right),$$

with an <u>arbitrary</u> exponent γ, giving $\beta = (1 + \gamma)/2$. This exponent would have been undetermined and would have to be found by other means. This leads to the following rule of thumb for the classification of self-similarity (Barenblatt):

(a) Self-similarity of the first kind: the problem is *regular* as function of the external parameters δ, L. The exponents are *rational*, determined by dimensional analysis, or symmetry, see below.

(b) Self-similarity of the second kind: the problem is *singular* as function of δ and L, the exponents are in general *irrational*, determined by a non-linear eigenvalue problem, see below.

Remark 2: The 'dynamical system' description. Assume for simplicity that $u = u(r, t)$ and put $\xi = r/\sqrt{t\kappa}$ as before, as well as <u>logarithmic time</u> $s = \ln t$. Put

$$u(r, t) =: \frac{1}{t\kappa} W(\xi, s), \tag{2.9}$$

so (2.1) can then be transformed into

$$\frac{\partial u}{\partial t} = -\frac{W}{t^2\kappa} - \frac{1}{2t^2\kappa}\xi W_\xi + \frac{W_s}{t^2} = \frac{1}{\kappa t^2}\frac{1}{\xi}(\xi W_\xi)_\xi \tag{2.10}$$
$$\to \frac{\partial W}{\partial s} = \frac{1}{\kappa}\left[W + \frac{\xi W_\xi}{2} + \frac{(\xi W_\xi)_\xi}{\xi}\right].$$

Thus a *fixed point* $\partial W/\partial s = 0$ of this dynamical system corresponds to the similarity solution (2.5), as it is equivalent to the similarity equation (2.4). The study of (2.10) is an extremely useful tool to study the *approach* to a similarity solution as well as its stability. In addition, the dynamical system (2.10) is the starting point to go beyond fixed-point behavior. For example, in some cases (2.10) has solutions corresponding to periodic orbits, which corresponds to a generalization of simple similarity solutions such as (2.8).

The heat equation (2.1) can of course be solved generally, so the approach to the similarity solution can be demonstrated explicitly, using Green functions. Namely, it is easy to verify that for the special initial condition $u(\mathbf{x},0) = Q\delta(\mathbf{x})$ the solution of (2.1) is (2.5). This must indeed be so, since (2.5) was found from the assumption $\delta \to 0$, which corresponds to an initial δ-function of zero extension.

As usual, a general initial condition can be written as a superposition of δ-functions:

$$u(\boldsymbol{x},0) = \int_{\mathbb{R}^2} \delta(\boldsymbol{x} - \boldsymbol{x}_0)u(\boldsymbol{x}_0,0)d^2x_0.$$

Thus from the superposition principle:

$$u(\boldsymbol{x},t) = \int_{\mathbb{R}^2} \frac{u(\boldsymbol{x}_0,0)}{4\pi\kappa t}e^{-\frac{|\boldsymbol{x}-\boldsymbol{x}_0|^2}{4\kappa t}}d^2x_0,$$

which is formally the most general solution.

To show that (2.11) always gives (2.5) in the limit $\delta \to 0$, we use a Taylor expansion of the kernel around $\boldsymbol{x}$, which is known as a multipole expansion in this context. Namely, let us assume that the initial blob is contained inside a ball of radius δ. The Taylor expansion is

$$e^{-\frac{|\boldsymbol{x}-\boldsymbol{x}_0|}{4\kappa t}} = e^{-\frac{|\boldsymbol{x}|}{4\kappa t}} + \frac{\boldsymbol{x}_0 \cdot \boldsymbol{x}}{2\kappa t}e^{-\frac{|\boldsymbol{x}|^2}{4\kappa t}} + O(\delta^2),$$

which gives

$$u(\boldsymbol{x},t) = \underbrace{\frac{Q}{4\pi\kappa t}e^{-|\boldsymbol{x}|/4\kappa t^2}}_{u_{\text{sim}}} + \frac{1}{4\pi\kappa t}\underbrace{\int u(\boldsymbol{x}_0,0)\boldsymbol{x}_0 d^2x_0}_{\boldsymbol{p}} \cdot \boldsymbol{x}\frac{e^{-|\boldsymbol{x}|^2/4\kappa t}}{2\kappa t} + O(\delta^2)$$

$$= u_{\text{sim}}(\boldsymbol{x},t) + \underbrace{\frac{\boldsymbol{p}\boldsymbol{x}}{8\pi\kappa^2 t^2}}_{}e^{-|\boldsymbol{x}|^2/4\kappa t} + O(\delta^2) = u_{\text{sim}}(\boldsymbol{x},t) + O\left(\frac{\delta}{\kappa^2 t^{3/2}},\delta^2\right).$$

$$|\boldsymbol{x}| \approx \sqrt{t},\, \boldsymbol{p} \approx \delta$$

Thus we can not only show that the solution converges to (2.5), but we can also compute corrections showing how this limit is approached.

3 The Power of Scaling

Physically, the spreading problem is characterized by the absence of a particular length scale; $\delta << R << L$, so R varies by many orders of magnitude.

Crucial Properties of scaling laws are:

(a) *universality* (independence of initial condition). n.b.: in the linear case the solution cannot be completely universal, as it is still depends on Q, set by the initial condition.

(b) *scaling form* $u = t^\alpha G(\frac{r}{t^\beta})$.

(c) *power laws*, either (i) rational (first kind) or (ii) irrational (second kind).

Figure 2. A naturally grown snowflake, from Libbrecht and Rasmussen (2003). It exhibits the six-fold symmetry of an ice crystal on a macroscopic scale.

But beware: there are examples where one or all of these properties break down: one often need to understand more about the problem. A particular case where things are more complicated is:

Example 1 *snowflakes*, see Figure 2. Not only does the form of the snowflake seem to depend on initial condition, it even depends on the (sixfold) symmetry of the crystal, although lattice constant $\simeq$ 1nm $<<<$ R! The reason must lie in instabilities, which cause small perturbations to get amplified, as it often occurs in far-from-equilibrium processes. Implicit in our analysis of (2.1) was the fact that certain microscopic parameters can be eliminated from the problem, an argument that cannot readily be made for the snowflake. However, an example where simple dimensional analysis provides deep insight into an extremely complicated problem is:

Example 2 *turbulent diffusion*, see Figure 3. The amount of stirring (turbulence) is characterized by the energy input ε per time and mass

$$\varepsilon = \frac{\text{energy}}{\text{time}} = \frac{\text{power}}{\text{mass}}, \quad [\varepsilon] = \frac{\text{cm}^2}{\text{s}^2 \cdot \text{s}} = \frac{\text{cm}^2}{\text{s}^3}.$$

We have

$$R(t) = f(t, \varepsilon, \delta, \nu),$$

Figure 3. Smoke spreading in a turbulent environment.

where δ is the initial size and ν the kinematic viscosity of the fluid in which the spreading occurs, $[\nu] = \mathrm{cm}^2/\mathrm{s}$.

If the transport is governed by turbulent motion alone, independent of fluid viscosity or initial condition, we have $R(t) = f(t, \varepsilon)$. The only way to combine ε and t to get a length is $[\varepsilon t^3] = \mathrm{cm}^2$, thus $R = A(\varepsilon t^3)^{1/2}$, much *faster* than $t^{1/2}$, i.e. by diffusion. **Diversion**: the idea of *dimensional analysis* is that no law of nature can depend on the units we give to the fundamental quantities of mass, length, time, and charge. Mathematically, this leads to a certain invariance under scale transformations (for details, see Birkhoff (1950)). More formally:

(i) Assume that there are n fundamental units $q_i, i = 1 \ldots n$, which can be changed by multiplying with a scale factor $\alpha_i : T(q_i) = \alpha_i q_i$.

(ii) The 'derived' quantities Q_j (such as density, velocity etc) then transform like $T(Q_j) = Q_j \Pi_{i=1}^n \alpha_i^{a_{ji}}$. The exponents a_{ji} are called the dimensions of Q_j. Now any physical law $f(Q_1, \ldots Q_k) = 0$ must be <u>unit free</u>, i.e. invariant under all transformations T.

Example 3 *Navier-Stokes equation*

$$\frac{\partial \boldsymbol{u}}{\partial t} + (\boldsymbol{u}\boldsymbol{\nabla})\boldsymbol{u} = -\frac{1}{\rho}\boldsymbol{\nabla}p + \nu\Delta\boldsymbol{u} \quad (NS) \tag{3.1}$$

All terms transform like $T(\partial \boldsymbol{u}/\partial t) = \mathrm{cm}/\mathrm{s}^2$, thus $T(NS) = NS$. This means all physical laws derived from (NS) must have the same invariance.

The statement that permits to simplify $f(Q_1, \ldots, Q_k) = 0$ given that $Tf = 0$ is known as the 'Buckingham Π-Theorem'. To formulate it, we have to make a technical assumption. Let $m \leqslant n$ be the rank of the a_{ij}- matrix defined by $T(Q_j)$, $j = 1, \ldots k$.

Then $f(Q_1, \ldots, Q_k) = 0$ is *equivalent* to $\phi(\Pi_1, \ldots, \Pi_{k-m}) = 0$, where the Π_ℓ are dimensionless. Thus one obtains a reduction of the number of variables by m without

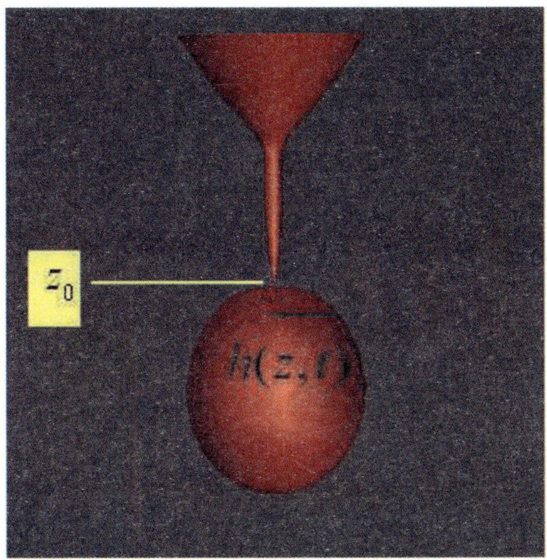

Figure 4. A drop separates from a faucet. At a time t_0 and position z_0 the radius of the fluid column goes to zero.

loss of information! The Π_ℓ correspond exactly to the expression $r/\sqrt{\kappa t}$ above and thus help to find power laws.

Applications include all systems with many length scales:
  (a) continuum mechanics: singularities (see below), turbulence, explosions, cracks
  (b) critical phenomena (systems near phase transitions), where correlation length $\to \infty$
  (c) elementary particles (where the energy becomes very large)
  (d) cosmology: black holes, etc.

Generalizations of scaling behavior are logarithms (breaking of scale invariance, see below), and *discrete* self-similarity, where solutions are only selfsimilar on a discrete set.

4 Drop Pinch-off: A Non-Linear Example

A drop falls from a faucet of radius r_0 and separates: this must be a singularity of the equations of motion. One way of seeing this is that after the singularity (separation) the drop lives on as an independent entity. Thus the mathematical problem changes its character at the singularity: what used to be one single equation describing a single drop of fluid, has now split into two independent equations. Another way of looking at it is that the radius of the fluid neck going to zero locally corresponds to the production of arbitrarily small scales in finite time, and thus to a singularity, i.e. a place where functions are no longer smooth. This will be evident from the similarity description given below.

How does the profile look near the breakup point z_0, t_0? The external parameters are ρ (density), ν (kinematic viscosity), γ (surface tension), g (acceleration of gravity), and r_0 (nozzle radius). We are interested in the limit $\Delta t = t_0 - t$ and $\Delta z = z - z_0$ small.

Thus for $\Delta t \to 0$, a typical length scale characteristic for breakup is $\ell_t = \sqrt{\nu \Delta t}$, keeping in mind that

$$[\nu] = \frac{\text{cm}^2}{s}, [\gamma/\rho] = \frac{\text{cm}^3}{s^2}, [g] = \frac{\text{cm}}{s^2}.$$

Thus using dimensional analysis,

$$h = f(\Delta z, \Delta t, s, \nu, \gamma, g, r_0)$$

can be reduced to

$$h = \ell_t \overline{f}\left(\frac{\Delta z}{\ell_t}, \frac{\ell_\nu}{\ell_t}, \frac{\ell_c}{\ell_t}, \frac{r_0}{\ell_t}\right), \tag{4.1}$$

where $\ell_\nu = \nu^2 \gamma/\rho$ is an intrinsic viscous length scale, and $\ell_c = \sqrt{\gamma/\rho g}$ is the (macroscopic) capillary length (typically 1 mm).

As argued before, in the limit $\ell_t \to 0$ the scales ℓ_c and r_0 shouldn't matter, so $\overline{f}$ should have a finite limit for $\ell_c/\ell_t \to \infty, r/\ell_t \to \infty$. But one still has

$$h = \ell_t \overline{f}\left(\frac{\Delta z}{\ell_t}, \frac{\ell_\nu}{\ell_t}\right), \quad v = \overline{g}\left(\frac{\Delta z}{\ell_t}, \frac{\ell_\nu}{\ell_t}\right),$$

where $v(z,t)$ is the average velocity in the z-direction.

Thus a (possible) similarity form cannot be found from dimensional analysis alone. However for the special case $\nu = 0$ (inviscid fluid), progress is possible. For example, for water $\ell_\nu \approx 10$ nm, so as long as $h_{\min} >> \ell_\nu$ we hope to effectively have $\ell_\nu = 0$. Assuming that this limit is indeed regular, we can redo our dimensional analysis, using $\ell_{in} = (\tilde{\gamma} \Delta t^2)^{1/3}$ as our length scale noting that $[\tilde{\gamma}] = [\gamma/\rho] = \text{cm}^3/s^2$. The only possible scaling form is then

$$h = \ell_{in} H_{in}\left(\frac{\Delta z}{\ell_{in}}\right), \qquad v = \frac{\ell_{in}}{\Delta t} V_{in}\left(\frac{\Delta z}{\ell_{in}}\right). \tag{4.2}$$

Some remarkable results follow:

$$h_{\min} = a_{in}(\gamma/\rho \Delta t^2)^{1/3}, \text{where} \quad a_{in} \approx 0.7 \tag{4.3}$$

from experiment and numerical simulations. Likewise, the entire scaling function H_{in}, which gives the form of the neck near pinch-off, is completely universal and can in principle be calculated!

In the more general case of finite ℓ_ν, one has to look at the equation underlying the dynamics. With some simplifications, they are

$$\frac{\partial h^2}{\partial t} + (h^2 v)_z = 0 \qquad \text{(mass conservation)}, \tag{4.4}$$

$$\frac{\partial v}{\partial t} + v v_z = -\tilde{\gamma}\left(\frac{1}{h}\right)_z + 3\nu \frac{(v_z h^2)_z}{h^2} + g \qquad \text{(force balance)}. \tag{4.5}$$

This set of equations has a special symmetry: if one substitutes $\tilde{h} = h/\tilde{\gamma}$, then ν and g are the only parameters left:

$$\frac{\partial \tilde{h}^2}{\partial t} + (\tilde{h}^2 v)_z = 0$$
$$\frac{\partial v}{\partial t} + v v_z = -\left(\frac{1}{\tilde{h}}\right)_z + 3\nu \frac{(v_z \tilde{h}^2)_z}{\tilde{h}^2} + g.$$

Thus if one writes the dimensionally correct expressions for $\tilde{h}$, which has dimensions $[\tilde{h}] = \mathrm{s}^2/\mathrm{cm}^2$, then $\ell_\nu = \nu^2/\tilde{\gamma}$ cannot appear, since it contains $\tilde{\gamma}$. But this means (4.2) can in fact be written as

$$\tilde{h} = \frac{(\Delta t)^2}{\ell_t^2} H\left(\frac{\Delta z}{\ell_t}\right), \quad v = \frac{\ell_t}{\Delta t} V\left(\frac{\Delta z}{\ell_t}\right).$$

Since $\tilde{\gamma}$ is just a constant, the similarity forms for the original variables h, v are

$$h(z,t) = t'\phi\left(\frac{z'}{t'^{1/2}}\right), \quad v(z,t) = t'^{-1/2}\psi\left(\frac{z'}{t'^{1/2}}\right), \tag{4.6}$$

where $z' = \Delta z/\ell_\nu, t' = \Delta t/t_\nu$ are convenient dimensionless length and time scales and $t_\nu = \nu^3 \rho^2/\gamma^2$. Again, it is implicit in the analysis that ϕ, ψ are universal. For example,

$$h_{\min} = a\frac{\gamma}{\rho\nu}\Delta t, \quad a \approx 0,03. \tag{4.7}$$

To compute ϕ, ψ, one inserts the similarity form into (4.4), (4.5). Namely, it is easily confirmed that in the force balance of (4.5)

$$\frac{\partial v}{\partial t} \sim v v_z \sim \left(\frac{1}{h}\right)_z \sim \frac{(v_z h^2)_z}{h^2} \sim \Delta t^{-3/2},$$

i.e. these terms blow up with the same power for $\Delta t \to 0$. By contrast, in (4.5), the contribution from gravity scales like $g \sim (\Delta t)^0$, and will drop out in the limit $\Delta t \to 0$. This is the important concept of *dominant balance*, which can also be used to show that <u>all terms</u> present in the hydrodynamic description of pinch-off are subdominant relative to the ones kept in the 'slender-jet' description (4.4), (4.5). The dominant balance means that gravity is not important for pinch-off, as was implicit in our earlier assumption that ℓ_c drops out from (4.1) in the limit $\Delta t \to 0$.

The similarity equations for ϕ, ψ, as obtained from (4.4) and (4.5), are

$$-2\phi^2 + \xi\phi\phi' + (\phi^2\psi)' = 0$$
$$\frac{\psi}{2} + \frac{\xi\psi'}{2} + \psi\psi' = -\left(\frac{1}{\phi}\right)' + 3\frac{(\psi'\phi^2)'}{\phi^2},$$

where $\xi = z'/t'^{1/2}$. This is an O.D.E. to be solved for ϕ, ψ; it can be solved uniquely using proper boundary constrictions, which we do not discuss here.

Finally, it is natural to look at the other limiting case where $\ell_\nu >> r_0$, much larger then all scales in the problem. Formally, this is best incorporated by putting $\rho = 0$, so there is no *inertia* in the problem. This leaves γ and $\eta = \nu \cdot \rho$ as the only parameters. Now $[\gamma/\eta] = \text{cm/s}$ and thus $\ell_{vis} = v_\eta \Delta t$ $(v_\eta = \gamma/\eta)$ is a convenient local length scale. Following the same logic as before one obtains:

$$h = \ell_{vis} H_{vis}\left(\frac{\Delta z}{\ell_{vis}}\right), \quad v = \frac{\ell_{vis}}{\Delta t} V_{vis}\left(\frac{\Delta z}{\ell_{vis}}\right).$$

All seems well, but the result is wrong!

The reason is that the limit $r_0/\ell_{vis} \to \infty$ is *not regular*, signaling self-similarity of the second kind. Namely, for $\rho = 0$ (4.5) reads

$$v_\eta \left(\frac{1}{h}\right)_z = 3\frac{(v_z h^2)_z}{h^2}\Big| \cdot h^2 \qquad \text{and so}$$

$$-h = \frac{3}{v_\eta} v_z h^2 - C(t), \tag{4.8}$$

or

$$\frac{3v(z)}{v_\eta} = -\int_0^z \left(\frac{C'}{h^2} + \frac{1}{h}\right) dz.$$

Thus v is determined by h alone!

The equations (4.4), (4.8) remain invariant under the transformation $z = a\tilde{z}$, $v = a\tilde{v}$, hence H_{vis}, V_{vis} cannot possibly be universal! Instead, the local solution must depend in some way on initial conditions, i.e. r_0. Let us assume that solutions remain self-similar, but depend on both combinations $\Delta z/\ell_{vis}$ and r_0/ℓ_{vis}. Then using the invariance under the above transformation we obtain

$$h = \ell_{vis} H_{vis}\left(\frac{\Delta z}{\ell_{vis}}\left(\frac{\ell_{vis}}{r_0}\right)^{1-\beta}\right), \quad v = u_\eta \left(\frac{\ell_{vis}}{r_0}\right)^{\beta-1} V_{vis}\left(\frac{\Delta z}{\ell_{vis}}\left(\frac{\ell_{vis}}{r_0}\right)^{1-\beta}\right). \tag{4.9}$$

To find the scaling exponent β, dimensional analysis evidently is not enough. One really has to use the (viscous) equation (4.9) together with (4.4).

To solve this viscous system of equations, we use a clever transformation, employing a particle marker as the spatial variable (Lagrangian transformation). Let α mark a fluid volume in the form of a slice $h^2(\alpha t)dz = d\alpha$. Then $z_\alpha = 1/h^2$ and $z_t = v$ (since α marks a material object). Here $z(\alpha, t)$ is the position of the volume α at time t. The first transformation incorporates volume conservation. Equation (4.8) transforms to (using $\partial_z = z_\alpha^{-1}\partial_\alpha \to v_z = z_\alpha^{-1}\partial_\alpha z_t$)

$$\frac{1}{z_\alpha^{1/2}} + \frac{3}{v_\eta}\frac{z_\alpha t}{z_\alpha^2} = C(t), \quad \text{which} \quad \text{with}$$

$$H(\alpha, t) = \frac{1}{\sqrt{z_\alpha}} \quad \text{is} \quad H - \frac{6}{v_\eta}H_t H = C(t). \tag{4.10}$$

The last equation no longer contains a spatial derivative! Introducing the dimensionless variables $t' = v_\eta \Delta t / r_0$, $\alpha' = \alpha / r_0^3$, we look for the similarity description

$$H = t' \chi \left(\frac{\alpha'}{t'^{2+\beta}} \right), \qquad C = t' \overline{C}, \tag{4.11}$$

where $\zeta = \alpha'/t'^{2+\beta}$.

The similarity equation is then

$$\overline{C} = \chi + 6\chi^2 - 6(2+\beta)\zeta\chi\chi'. \tag{4.12}$$

The constant can be determined using a constraint implicit in $z_\alpha = 1/H^2$, and hence $1/\chi^2 = F_\zeta$, where $F(\zeta)$ is some function. Dividing by χ^4:

$$\overline{C}/\chi^4 = 1/\chi^3 + \underbrace{6/\chi^2}_{6F_\zeta} - 6(2+\beta)\underbrace{\zeta\chi'/\chi^3}_{-\zeta F_{\zeta\zeta}/2}$$

$$\underbrace{\qquad\qquad\qquad\qquad}_{[F + 3(2+\beta)(\zeta F_\zeta - F)]_\zeta}$$

and thus

$$\overline{C} \int_{-\infty}^{\infty} \frac{d\zeta}{\chi^4} = \int_{-\infty}^{\infty} \frac{d\zeta}{\chi^3}. \tag{4.13}$$

Integrating (4.13) we get

$$\frac{1}{6(2+\beta)} \ln \left| \frac{\zeta}{\overline{\zeta}} \right| = \int_{\overline{\chi}}^{\chi} \frac{\zeta \, d\zeta}{6\zeta^2 + \zeta - \overline{C}}, \quad \overline{\chi} = \chi(\overline{\zeta}).$$

Locally (near the minimum) χ looks like $\chi = \chi_0 + \zeta^2 + O(\zeta^4) + \cdots$.

Here we used the fact that the ζ-scale is arbitrary, so the quadratic coefficient was normalized to unity. Performing the integral gives

$$\left| \frac{\zeta}{\overline{\zeta}} \right|^{\frac{2}{2+\beta}} = \left| \frac{6\chi^2 - \chi - \overline{C}}{6\overline{\chi}^2 - \overline{\chi} - \overline{C}} \left(\frac{(\Gamma + 1 + 12\chi)(\Gamma - 1 - 12\overline{\chi})}{(\Gamma - 1 - 12\chi)(\Gamma + 1 + 12\overline{\chi})} \right)^{\frac{1}{\Gamma}} \right|, \tag{4.14}$$

where $\Gamma = \sqrt{1 + 24\overline{C}}$. Since $\overline{\chi} \to \chi_0$ as $\overline{\zeta} \to 0$ one must have $6\chi_0^2 - \chi_0 - \overline{C} = 0$, and since powers of ζ must match for $\zeta \to 0$, we get $-\frac{2}{2+\beta} = -2 + \frac{2}{\Gamma}$. Combining both equations results in

$$\chi_0 = \frac{1}{12(1+\beta)}, \qquad \overline{C} = \frac{3+2\beta}{24(1+\beta)^2}. \tag{4.15}$$

Now (4.13) can be understood as an equation for β! By changing the variable of integration from ζ to χ, the integrals can be computed in the following three steps:

(i) Equation (4.12) can be solved to give

$$\zeta = \left[\frac{6(1+\beta)}{2+\beta} \left(\chi + \frac{3+2\beta}{12(1+\beta)} \right) \right]^{\frac{3+2\beta}{2}} (\chi - \chi_0) \tag{4.16}$$

(ii) One has

$$\int_{-\infty}^{\infty} \frac{d\zeta}{\chi^i} = 2 \int_{\chi_0}^{\infty} \frac{d\chi}{\chi^i \chi'}, \text{ where } \chi' = \frac{6\chi^2 + \chi - \overline{C}}{6(2+\beta)\chi\zeta} \tag{4.17}$$

from (4.12), and ζ is expressible through χ by (4.16).

(iii) Evaluating both sides of (4.13) as integrals over χ gives

$$\frac{(1-\beta)(3+2\beta)}{(1+\beta)(3-2\beta)} = \frac{F(-\frac{1}{2}-\beta, 1-\beta; 3/2-\beta; 3-2\beta)}{F(-\frac{1}{2}-\beta, 2-\beta; 5/2-\beta; -3-2\beta)} \tag{4.18}$$

where $F(a, b; c; d)$ is Gauss's hypergeometric function (Abramowitz and Stegan, 1970). Solving this last equation gives $\beta = 0,1748717\cdots$.

Thus the nonlinear eigenvalue problem (4.12), (4.13) indeed leads to a nontrivial, irrational exponent which is typical for self-similarity of the second kind. One then finds $\chi_0 \approx 0,071$, thus

$$h_{\min} = a_{vis}\frac{\gamma}{\rho\nu}\Delta t, \ a_{vis} = 0,0709. \tag{4.19}$$

Again, the minimum of H_{vis} is universal, however the entire function is not, its width depends on initial conditions, as shown earlier.

5 Logarithms: Lavalamp Dynamics

A novel phenomenon occurs if dripping takes place inside an ambient fluid: the simplest case is that of the viscosities of both fluids being equal. The interface then behaves like an elastic membrane of surface tension γ. This means each point of the interface exerts a point force $\mathbf{n}\cdot$ curvature $\approx \mathbf{n}/h$ if the profile is slender. The total velocity can be written as an *integral* over all point forces:

$$v(z) = \frac{1}{4} \int_{\text{drop}} \frac{h_{z'}(z')}{\sqrt{h^2(z') + (z-z')^2}} dz',$$

having simplified the expression by assuming $h_z(z) << 1$.

This is to be coupled to the usual mass balance (4.4). Dimensionally, one would once more expect a local solution of the form

$$h(z, t) = \ell_{\text{vis}}H_{\text{out}}\left(\frac{\Delta z}{\ell_{\text{vis}}}\right), \quad \ell_{\text{vis}} = v_\eta \Delta t.$$

Now as $\ell_{\text{vis}} \to 0$ and Δz is finite, one expects $h(\Delta z, t_0)$ to remain finite. But this means H_{out} must be linear at infinity

$$H_{\text{out}}(\xi) \to \begin{cases} H_+\xi & \xi \to \infty \\ & \\ H_-\xi & \xi \to -\infty. \end{cases} \qquad \xi = \frac{\Delta z}{\ell_{\text{vis}}}$$

In similarity variables, (5.1) has the form

$$V_{\text{out}}(\xi) = \frac{1}{4} \int_{-z_b/\ell_{\text{vis}}}^{z_b/\ell_{\text{vis}}} \underbrace{\frac{H'_{\text{out}}(\xi')}{\sqrt{H^2_{out} + (\xi - \xi')^2}}}_{\text{arg}} d\xi'.$$

But if H_{out} is linear, the argument behaves like

$$\text{arg}(\xi') \rightarrow \begin{cases} (H_+/\sqrt{H^2_+ + 1})\xi, & \xi \rightarrow \infty \\ (H_-/\sqrt{H^2_- + 1})\xi, & \xi \rightarrow -\infty \end{cases}$$

which results in a <u>divergent</u> integral!

This can be <u>fixed</u> using a simple trick:

$$h = \ell_{\text{vis}} H_{\text{out}}(\xi), \quad \xi = \frac{\Delta z}{\ell_{\text{vis}}} + b \ln \Delta t, \tag{5.1}$$

i.e. moving the z-axis at a speed which depends logarithmically on time! With this modification, the mass balance becomes

$$-H_{\text{out}} + H'_{\text{out}}(\xi + V_{\text{out}} - b \ln \Delta t) = H_{out} V'_{out}/2.$$

Now we choose b such that the logarithmic singularity cancels, namely

$$\frac{1}{4} \int_{-z_b/\ell_{\text{vis}}}^{z_b/\ell_{\text{vis}}} \frac{H'_{out}(\xi')}{\sqrt{H^2_{out} + (\xi - \xi')^2}} d\xi' - b \ln(\ell_{\text{vis}})$$

finite for $\ell_{\text{vis}} \rightarrow 0$. This is achieved by putting

$$b = \frac{1}{4} \left[-\frac{H_+}{\sqrt{H^2_+ + 1}} + \frac{H_-}{\sqrt{H^2_- + 1}} \right].$$

Thus defining

$$V_{\text{fin}}(\xi) = \lim_{\Lambda \rightarrow \infty} \frac{1}{4} \int_{-\Lambda}^{\Lambda} \frac{H'_{\text{out}}}{\sqrt{H^2_{\text{out}} + (\xi - \xi')^2}} d\xi' + b \ln \Lambda \tag{5.2}$$

the similarity equation

$$-H_{\text{out}} + H'_{\text{out}}(\xi + V_{\text{fin}} - \xi_0) = H_{out} V'_{\text{fin}}/2 \tag{5.3}$$

is finite, and ξ_0 is an arbitrary constant.

The numerical solution of this integro-differential equation gives

$$h_{\min} = a_{\text{out}} v_\eta \Delta t, \quad \text{where} \quad a_{\text{out}} = 0,0335 \quad \text{and} \quad H_+ = 4,81; H_- = -0,105. \tag{5.4}$$

These constants are universal, the whole similarity profile H_{out}, however, is not, for the following reason: whereas in previous examples the singular motion was self-similar,

$$h(z,t) = \ell_{\mathrm{vis}} H_{\mathrm{out}} \left(\frac{\Delta z}{\ell_{\mathrm{vis}}} + b \ln \Delta t \right) \tag{5.5}$$

is not: a change in ℓ_{vis} cannot be absorbed into rescaling of the axes.

6 Nanojets: Fluctuations

On small scales, thermal motion becomes ever more important. Fluid elements behave like Brownian particles, and lead to increasing fluctuations. To estimate fluctuations of an interface, one constructs the thermal length scale $\ell_T = \sqrt{k_B T / \gamma}$, comparing thermal and surface tension energies; it is typically 1nm. This means a liquid bridge 'wobbles' increasingly as it pinches off.

This behavior is indeed seen in simulations (microscopic 'MD' simulations of individual atoms (Moseler and Landman (2000))) and in experiments. The experiments (Aarts et al. (2004)) use a colloid-polymer mixture, which separates into two phases, with very small ($\gamma \approx 0, 2\mu\mathrm{N/m}$) surface tension between them. Thus ℓ_T is $\approx 10\mu m$, and fluctuations are macroscopically observable.

In particular, one important observation is made: instead of $h_{\min} \propto \Delta t$, pinch-off occurs much more abruptly: typically $h_{\min} \propto \Delta t^{0.4}$. How do fluctuations succeed in driving the system toward a singularity more rapidly? To answer this question, fluctuations must be added to the dynamics. To avoid complications, we introduce a radical simplification: the entire bridge is treated as if it were just a single particle of mass $m_{\mathrm{eff}} = m(h)$. Of course, this mass will be smaller if the bridge radius is small, see Figure 5.

The simplest possible set of equations is

$$\dot{h} = v \qquad m(h)\dot{v} = \xi, \tag{6.1}$$

where $\xi(t)$ is Gaussian white noise. This means a discrete version is:

$$P(\xi_j) = e^{-\xi_j^2/2}, < \xi_i \xi_j >= \delta_{ij}.$$

The probability of finding an entire sequence is thus

$$\omega\{\xi_i\} = \exp\left\{ -\frac{1}{2} \sum_i \xi_i^2 \right\}.$$

For a *continuous* path $h(t), v(t)$ we thus have the probability

$$\omega\{h(t), v(t)\} = e^{-S}, \qquad S = \frac{1}{2} \int (m\dot{v})^2 dt.$$

Formally, *any* path $h(t)$ is possible, but each is assigned a different probability. A way to characterize the *typical* behavior is to look for the most probable path. Such an optimal path is $h = h_0, v = 0$, since this makes $S = 0$, which is an absolute minimum.

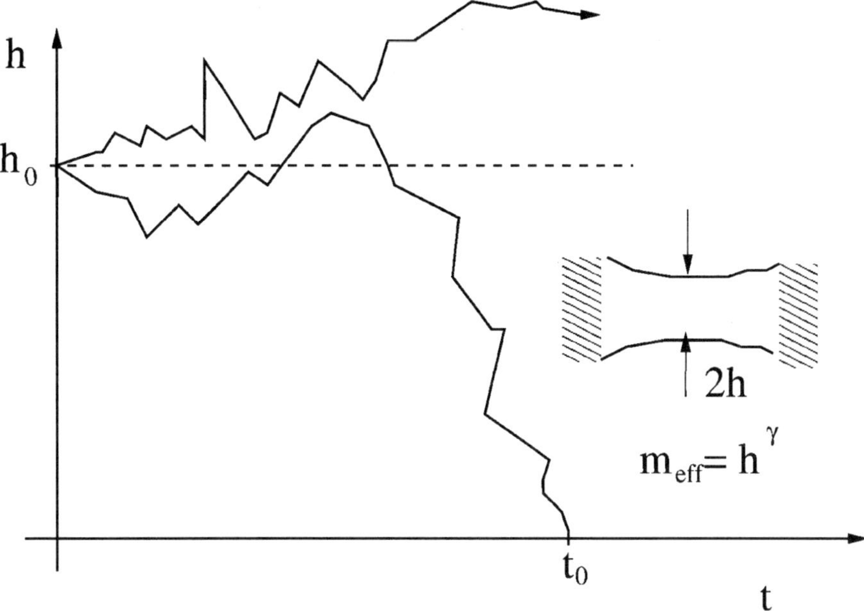

Figure 5. A very simple model of a "nanobridge". The only variable is the minimum radius h. We are only interested in paths which lead to pinch-off in some finite time t_0.

However, this ignores the fact that paths do not last forever, but rather end at different singular times t_0. It is thus a more well-defined question to fix t_0 and to ask for the most probable path that begins at $(0, h_0)$, and ends at $(t_0, 0)$! This corresponds to the classical Euler-Lagrange problem of finding the minimum of S with fixed endpoints. The equation $\dot{h} - v$ has to be built in as a constraint:

$$S = \int \frac{m^2(h)}{2}\dot{v}^2 + \tilde{h}(\dot{h} - v)dt.$$

The Euler-Lagrange equations read

$$\frac{d}{dt}\frac{\partial S}{\partial \dot{v}} - \frac{\partial S}{\partial v} = \frac{d}{dt}m^2\dot{v} + \tilde{h} = 0.$$

$$\uparrow \text{`force}'$$

$$\frac{d}{dt}\frac{\partial S}{\partial \dot{h}} - \frac{\partial S}{\partial h} = \dot{\tilde{h}} - \gamma h^{2\gamma-1}\dot{v}^2 = 0.$$

The intriguing feature is that an effective force $\tilde{h}$ has emerged, which could drive the bridge toward pinch-off.

It is convenient to pass to a 'Hamiltonian' description with momentum variables $\tilde{h} = \partial S/\partial \dot{h}$ and $\tilde{v} = \partial S/\partial \dot{v} = m^2\dot{v}$. This leads to the <u>four</u> equations

$$\dot{h} = v \quad \dot{\tilde{h}} = \gamma\tilde{v}^2/h^{2\gamma+1}$$
$$\dot{v} = \tilde{v}/h^{2\gamma} \quad \dot{\tilde{v}} = -\tilde{h}, \tag{6.2}$$

which are to be solved with *boundary values* at $t = 0$ and t_0. Instead of solving (6.2) completely, we are just interested in scaling solutions for $\Delta t = t_0 - t \to 0$:

$$
\begin{aligned}
h &= A_1 \Delta t^{\alpha_1} & \tilde{h} &= A_3 \Delta t^{\alpha_3} \\
v &= A_2 \Delta t^{\alpha_2} & \tilde{v} &= A_4 \Delta t^{\alpha_4}.
\end{aligned}
\tag{6.3}
$$

Inserting this into (6.2), there are non-trivial solutions for the set of exponents. In particular

$$
\alpha_1 = \frac{4}{2\gamma + 2}.
\tag{6.4}
$$

Thus indeed if $\gamma > 1$(the mass decreases sufficiently rapidly with h) $\alpha_1 < 1$, and pinch-off is faster than linear!

Bibliography

D.G.A.L. Aarts, M. Schmidt, and H.N.W. Lekkerkerker. Direct visual observation of thermal capillary waves. *Science*, 304:847–850, 2004.

M. Abramowitz and I.A. Stegan. *Handbook of Mathematical Functions*. Dover, 1970.

G.I. Barenblatt. *Scaling, Self-Similarity, and Intermediate Asymptotics*. Cambridge University Press, 1996.

G. Birkhoff. *Hydrodynamics: A Study in Logic, Fact, and Similitude*. Princeton University Press, 1950.

K. Libbrecht and P. Rasmussen. *The Snowflake: Winter's Secret Beauty*. Voyageur Press, 2003.

M. Moseler and U. Landman. Formation, stability and breakup of nanojets. *Science*, 289:1165–1169, 2000.

Three-Phases Capillarity

David Quéré

Laboratoire Hydrodynamique et Mécanique Physique, UMR CNRS, ESPCI, Paris, France.

Abstract We discuss a few situations where a liquid is brought into contact with a solid, with air around. Starting from the very classical Young equation, we present recent developments in the field, such as the possibility of generating spontaneous motions, or the wetting of textured surfaces. Then, we describe a few dynamic situations, in particular wicking, creeping and coating.

1 Introduction

A surface separating two different media has a specific energy, which arises from the contrast between the two media (Rowlinson and Widom, 1982). This energy is proportional to the surface area of the considered interface, and dominates gravitational energy, provided that the system is small enough. In many cases, the systems exhibit several or even many surfaces (which can be of the same kind, as in a foam, or different, as in wetting), so that the energy minimization leading to the minimum of the total surface energy of the system can lead to a large variety of situations.

Here we consider a canonical theme of this kind, namely a drop on an ideal solid, and develop several variations on this theme. In the first part we address static problems: we discuss the force acting on the liquid when deposited on a chemical discontinuity, and deduce a few situations of spontaneous motions; then, we summarize recent advances on the wetting on micro-structured surfaces. In the second part, we present dynamical problems, such as wicking, viscous resistance of a moving drop, and fluid coating. These different questions were selected for pedagogical reasons: they are all central, of practical importance, and can be viewed as classical issues in the field. We thus think that they deserve a brief description in a general book devoted to deformable interfaces. We favored a qualitative and intuitive approach, and indicated historical or recent papers in the references.

2 Statics of Wetting

2.1 Young Equation

When deposited on a solid, a small amount of liquid will either spread or form a little lens. As proposed by Marangoni, a convenient way to understand which of these possibilities will happen consists of comparing the surface energy of the dry solid, with the ones of the wet solid, as sketched in Figure 1 (de Gennes et al., 2004).

We denote as S the spreading parameter, which compares the surface energies per unit area of both these situations:

$$S = \gamma_{SA} - \gamma_{SL} - \gamma_{LA} \tag{2.1}$$

where γ designates the different surface tensions, and the subscripts refer to the phase in presence (S for solid, L for liquid and A for air). We later simply denote as γ the quantity γ_{LA}. The sign of S decides the wetting behavior. If S is positive, the system lowers its surface energy by being wet, corresponding to Figure 1b (complete wetting). In the opposite case ($S < 0$), a drop will only partially wet the solid.

Figure 1. The wetting behavior of a liquid on a solid can be predicted by comparing the surface energy of the dry solid (left) with the ones of a wet solid (right).

Partial wetting is the most common situation (complete wetting implies the replacement of one interface by two interfaces, which is generally less favorable). If small, the drop will form a spherical cap (which satisfies the equilibrium condition of a constant inner pressure $P = P_0 + 2\gamma/R$, denoting R as the radius of curvature of the drop), which meets the solid with an angle θ. If the solid is ideal, that is, homogeneous chemically and physically, this angle is unique, and independent of the drop size (Figure 2).

The question is of course to understand what fixes this angle, which quantifies the wettability of the solid. A good way to understand this problem consists of making a zoom in the vicinity of the contact line, where the three phases in presence meet. Let us consider a small displacement of the contact line, by a quantity dx (Figure 3). This increases the solid/liquid and liquid/vapour interfaces, and decreases the solid/air interface. (Due to volume conservation, it also decreases the angle θ, but the corrections this induces are of second order.) The corresponding variation of surface energy (per unit length of the contact line) can be written:

$$dE = (\gamma_{SA} - \gamma_{SL})dx + \gamma dx \cos\theta. \tag{2.2}$$

Equilibrium is given by the minimum of E ($dE = 0$), which yields the famous Young relation (Young, 1805):

$$\gamma \cos\theta = \gamma_{SA} - \gamma_{SL} \tag{2.3}$$

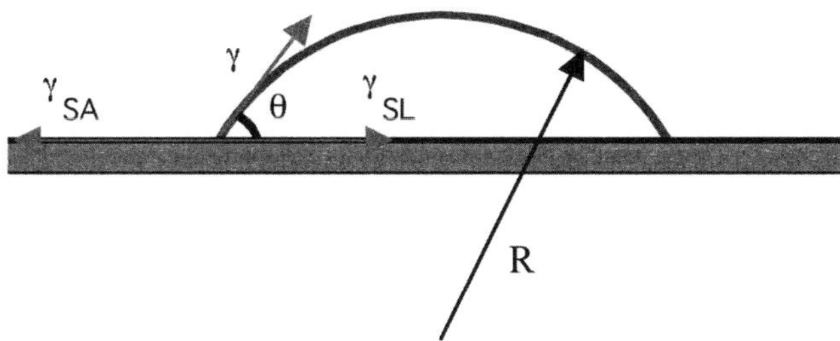

Figure 2. Partial wetting of a small drop on an ideal solid. Then, the drop meets the solid with a unique contact angle given by the Young equation.

which can also be expressed as a function of the spreading parameter S:

$$\cos\theta = 1 + \frac{S}{\gamma}. \tag{2.4}$$

This form allows us to check that indeed partial wetting corresponds to a negative spreading parameter; the wetting transition towards complete spreading occurs for $S = 0$, for which the contact angle θ vanishes.

2.2   Spontaneous Motions of Drops

These elementary notions allow us to understand remarkable behaviors such as the possible spontaneous displacement of a drop. The simplest situation is the one of a drop placed across a hydrophilic/hydrophobic frontier on a solid. [Such solids are easily prepared with a hydrophilic solid, as glass for water; then, half the solid is treated by dipping it in a solution containing fluorosilanes which covalently attach to glass, making it strongly hydrophobic.] We represent this situation in Figure 4, where solid 1 is hydrophobic (there, the drop would make an angle θ_1), and solid 2 is hydrophilic (there it would make an angle θ_2, with $\theta_2 < \theta_1$).

The conformation of the drop sketched in Figure 4 is not that obvious. The drop could choose to meet solid 1 with the angle θ_1, and solid 2 with the angle $\theta_2 > \theta_1$. As a consequence, the drop would be asymmetric, with a higher curvature (that is, pressure) on 1 than on 2: thus, it should move towards 2. The motion is indeed observed, but not the asymmetry in the profile: the pressure tends to be uniform in the liquid, which imposes a constant curvature, as represented in Figure 4. Hence, the drop meets the different solids with the same angle θ, which must be intermediate between θ_1 and θ_2.

We can then make a Young construction on each contact line. For the sake of simplicity, we consider here a two-dimensional drop, or rim, whose cross-section is pictured

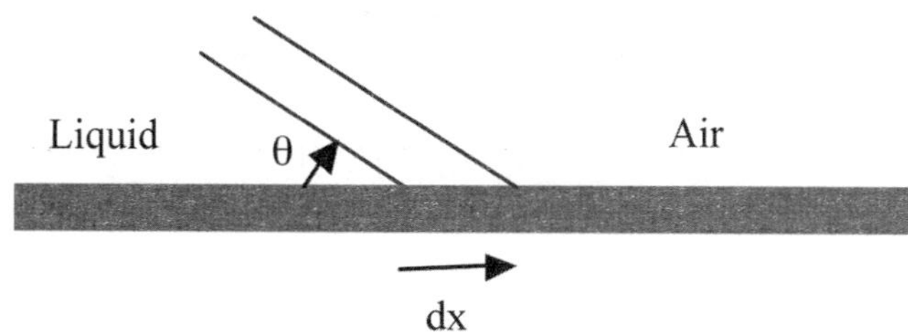

Figure 3. We consider here a displacement of the contact line by a quantity dx, and calculate in equation 2.2 the corresponding variation of surface energy.

in Figure 4. On the left, the angle is smaller than the equilibrium angle θ_1 (given by the Young relation $\gamma \cos \theta_1 = \gamma_{1A} - \gamma_{1L}$), and the force acting on the line projected on the horizontal is $\gamma_{1L} - \gamma_{1A} + \gamma \cos \theta$, i.e. $\gamma(\cos \theta - \cos \theta_1)$. Similarly, the horizontal projection of the force acting on line 2 is $\gamma_{2A} - \gamma_{2L} - \gamma \cos \theta$, that is $\gamma(\cos \theta_2 - \cos \theta)$. Adding these forces leads to a formula for the force F per unit length acting on the drop, and which is found to be independent of θ, but is a function of the contrast of wettability across the frontier:

$$F = \gamma(\cos \theta_2 - \cos \theta_1). \tag{2.5}$$

The situation considered here only induces a very local displacement, of the order of the drop size: once all of the drop is in the hydrophilic region, the motion stops. This can be useful for understanding local rearrangements on a heterogeneous substrate, but cannot provide extensive motions. It was thus proposed to create substrates decorated with a gradient of wettability, for which the drop is always in the situation depicted in Figure 4. Beautiful experiments were conducted on this subject by Chaudhury and coworkers (e.g. Daniel and Chaudhury (2002)) who observed drop motions driven by the gradient of wettability. In fact such motions occur even on slightly inclined plates. Indeed, tilting the substrate is a very convenient way of measuring the driving force: there is a critical angle for which the motion stops and from which the driving force is immediately deduced. The drops flattens as it progresses, because is goes to regions which are more and more hydrophilic. However, as pointed out by Chaudhury and co-workers, pinning of the line on defects opposes the motion, and because the force arising from pining will often be of the order of F, the drop can just get stuck on such surfaces. It was thus proposed to vibrate the substrate, which allows the contact line to depin, and favors considerably the motion.

Remarkable variations on this theme were proposed between 1995 and 2005. We understood that an asymmetric wettability promotes a motion, and different groups

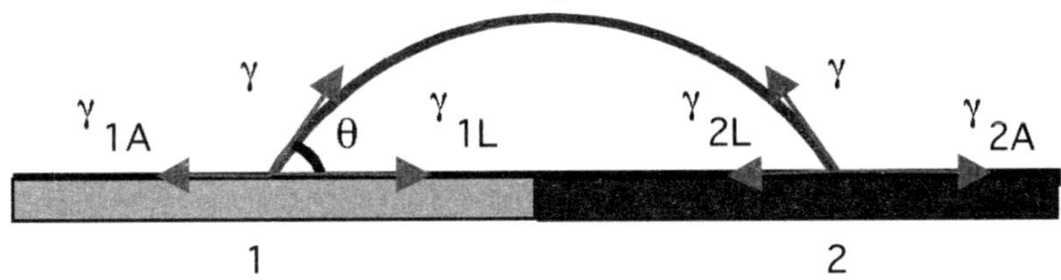

Figure 4. Drop placed across hydrophobic/hydrophilic surface. The drop will escape from the hydrophobic region and move towards the hydrophilic region (as expected from etymology).

thought about tricks for inducing such an asymmetry. A first series of achievements concern the solid itself. Apart from the natural idea of a gradient of wettability, it was proposed to set an asymmetric wettability using electrowetting: placing a drop between two electrodes modify its contact angle (whose value then results from a minimization of the surface energy plus the electrostatic energy): the higher the voltage, the smaller the angle (Lippman's law). Embedding electrodes inside the substrate, perpendicularly to the desired motion, indeed make a drop move by propagating a voltage along the electrodes. Similar effects were observed using temperature gradients, but we will not discuss this effect, which is more complex because of the variations of all the surface tensions with temperature (de Gennes et al., 2004). It was also cleverly proposed to vibrate the substrate (parallel to itself) with an asymmetric vibration: the balance of inertial forces acting on the drop leads to a motion (Daniel et al., 2005).

Restricting to chemical modifications, it was suggested that the asymmetry of wettability could be brought by the drop itself. A first experiment consists of dissolving a chemical compound in the drop, such as fluorosilanes in octane (Bain et al., 1994; Domingues Dos Santos and Ondarçuhu, 1995). Depositing a drop of this kind on a glass substrate leads to a spontaneous motion, where the trajectory is observed to follow a self-avoiding random walk (when the drop comes to a place where it passed, it bounces back). This system is sketched in Figure 5. As the drop is deposited, the fluorosilane starts to react with glass, making it less wettable for all liquids. Octane wets glass, but not if glass is covered by a monolayer of silanes. The place where the reaction starts is random, and may depend on the way the liquid is deposited, or on defects present on the solid surface. But as soon as the reaction starts, the drop tends to escape from the hydrophobic region it created, so that the motion persists. If the drop meets less wettable regions, it tends to avoid them, which leads to the random motion (more quantitative experiments were performed on tracks on which the trajectory is nicely linear). In particular, if the drop meets its own track, it bounces back: this self-motion just works once.

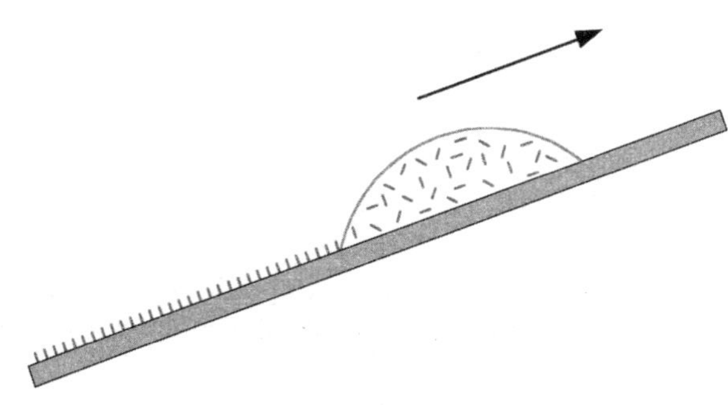

Figure 5. Self-running drop. The motion comes from a chemical reaction at the rear of the drop, which makes the solid at this place less wettable than at the front.

A different device was recently proposed, making it possible for the drop to cross its own path (Sumino et al., 2005). In this experiment, the solid (glass) is first covered by surfactant molecules, which are physi-sorbed on the substrate (with a much weaker chemical bound than above). These surfactants are provided by a bath of soapy water in which the glass is immersed. Then a drop of denser oil is deposited on the glass plate. This oil contains iodine, which tends to make a complex with the surfactant molecules, which are thus extracted from the surface instead of being deposited as in Figure 5. Hence, there is an asymmetry in the drop (provided that the extraction process starts at some point), which moves till all the iodine is complexed (which can be hours). Interestingly, once the drop has moved, surfactants adsorb again, so that the surface is renewed, allowing the drop to pass again at this place.

2.3   Wetting of Textured Surfaces

Another field which attracted a lot of interest during the last decade is the wetting of textured surfaces. Most of surfaces are rough, and it is interesting to understand how this modifies the wettability of these surfaces. Importantly, roughness does not only change the apparent contact angle, but it also allows an interval of angles instead of a unique one. The minimum and maximum angles are called the 'receding' and 'advancing angles', and the difference between both (often of the order of tens of degrees) is referred to as the 'contact angle hysteresis'.

Roughness can also be intentional. Starting from an ideal surface (contact angle θ, hysteresis $\Delta\theta = 0$), we shall discuss how the introduction of a well defined microtexture (as produced using microelectronics technologies) modifies the contact angle and the contact angle hysteresis. In addition, we shall show that such textures generate original wetting properties, which could not be achieved without them.

Shibuichi et al. (1996) conducted a remarkable series of experiments on this topic. Their results are displayed in Figure 6, where the value of the measured apparent contact angle θ^* on a rough surface is plotted as a function of θ, the Young contact angle determined on a flat surface of the same chemical composition. Both angles are expressed by their respective cosines.

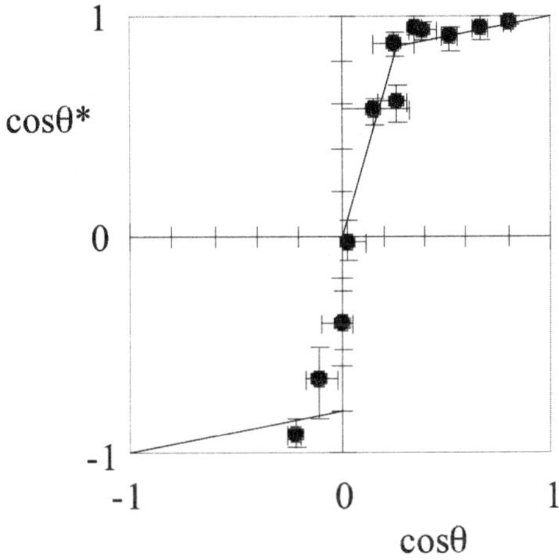

Figure 6. Contact angle on a rough surface as a function of the contact angle on a surface of the same chemical composition, yet flat (Shibuichi et al., 1996).

As a main result, the natural tendency of the solid is enhanced by the presence of a texture, which will make a hydrophilic material ($\theta < \pi/2$) more hydrophilic ($\theta^* < \theta$), and a hydrophobic material ($\theta > \pi/2$) more hydrophobic ($\theta^* > \theta$). In the latter case, solids can even become super-hydrophobic (θ^* larger than typically 160°), which is not possible using only chemical means (most hydrophobic flat surfaces have contact angles of the order of 120°).

A first idea for understanding these effects consists of assuming that the liquid still follows the solid surface when it is rough (Wenzel model). For deriving the contact angle, we consider again the displacement of the contact line as sketched in Figure 3, yet on a rough surface (we define a roughness factor r as the ratio of the actual surface area over the projected one). The apparent displacement of the line is dx, but the actual displacement is rdx, so that the corresponding change in surface energy can be written as (per unit length of the line):

$$dE = r(\gamma_{\mathrm{SL}} - \gamma_{\mathrm{SA}})dx + \gamma dx \cos\theta^* \qquad (2.6)$$

since the liquid/vapour interface is not affected by the solid roughness. E is minimal at equilibrium, which yields *Wenzel's relation* (Wenzel, 1936):

$$\cos\theta^* = r\cos\theta \qquad (2.7)$$

where θ is Young's angle (Equations (2.3) and (2.4)).

Equation (2.7) qualitatively explains the Shibuichi observations. The roughness factor being larger than unity, a hydrophilic surface is expected to be more hydrophilic if the surface is rough, and hydrophobicity will be similarly enhanced by the material roughness. However, a quantitative comparison is less favorable (Bico et al., 2002): in Figure 6, a linear relationship between $\cos\theta^*$ and $\cos\theta$ is only observed on the hydrophilic side (suggesting a very simple and cheap method for measuring a surface roughness), and only till some critical value of $\cos\theta$. For solids which are 'too hydrophilic', a different law is followed, which also seems to be linear in this plot, yet with a slope smaller than unity. In particular, contrasting with what is expected from equation (2.7), the modified apparent contact angle θ^* cannot be zero.

This second regime can be understood by supposing that the liquid is able to penetrate the network of cavities, as it could do in a porous medium (Figure 7). When a liquid invades a porous medium, the solid/air surface of the pores constituting the medium becomes replaced by a solid/liquid surface. This will occur if the corresponding surface tensions satisfy the condition $\gamma_{\mathrm{SL}} < \gamma_{\mathrm{SA}}$, which is equivalent, as shown by the Young relation, to a condition of hydrophilicity ($\theta < \pi/2$). Here (Figure 7), the propagation of a liquid film does not only imply to fill the cavities of the solid (as in classical wicking), but it also generates the creation of new liquid/vapour interfaces. The energy change related to a propagation of the film by a quantity dx can thus be written as $dE = (\gamma_{\mathrm{SL}} - \gamma_{\mathrm{SA}})(r - \phi_s)dx + \gamma(1 - \phi_s)dx$, where ϕ_s is the solid fraction remaining dry (i.e. the top of the posts in Figure 7). Then film propagation will occur if dE is negative. Together with Young's relation, this condition writes:

$$\theta < \theta_c, \quad \text{with} \quad \cos\theta_c = \frac{1 - \phi_{\mathrm{S}}}{r - \phi_{\mathrm{S}}}. \qquad (2.8)$$

Since this process is somehow intermediate between spreading and wicking, we logically find an intermediate criterion. Both spreading and wicking appear as limiting cases in equation (2.8). For a flat surface ($r \to 1$), we find $\theta_c = 0$ - as stressed in equation (2.4)

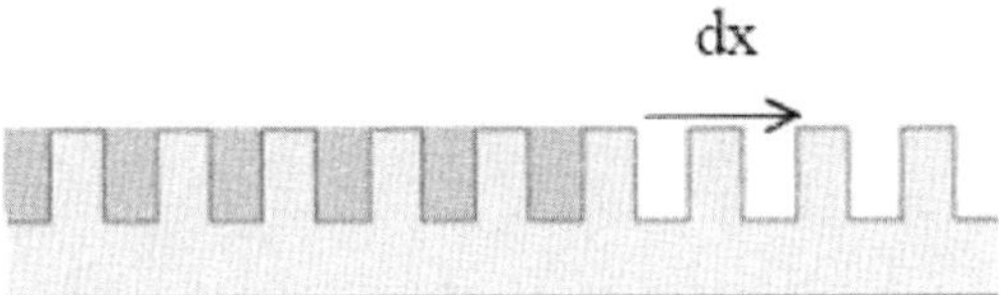

Figure 7. Liquid film (in grey) propagating in microtexture which consists of regularly spaced microposts.

- and spreading on a flat surface occurs if the contact angle vanishes. In a porous medium ($r \to \infty$), we have $\theta_c = \pi/2$, which is the classical condition for capillary rise. In our case ($r > 1$ and $\phi_s < 1$), equation (2.8) defines a critical contact angle between 0 and $\pi/2$. This implies that the ability of a textured surface to drive a liquid can be tuned by its design (Bico et al., 2002).

If equation (2.8) is satisfied, the contact angle has an expression different from that expected by equation (2.7). Then, a film propagates from the drop, which sits on a mixture of solid and liquid, such as sketched on the left of Figure 7. More generally, the contact angle on a surface constituted of different chemicals (yet flat) was addressed by Cassie and Baxter (1944), for a surface made of two species (of respective contact angles θ_1 and θ_2, and relative surface proportions ϕ_1 and ϕ_2).

A small displacement dx of the contact line implies, as sketched in Figure 8, a variation of surface energy $dE = \phi_1(\gamma_{SL} - \gamma_{SA})_1 dx + \phi_2(\gamma_{SL} - \gamma_{SA})_2 dx + \gamma dx \cos\theta$. Using the laws for both chemical species, and assuming that the energy is minimum, yields the apparent contact angle on such a surface:

$$\cos\theta^* = \phi_1 \cos\theta_1 + \phi_2 \cos\theta_2. \tag{2.9}$$

This formula first derived by Cassie and Baxter allows us to predict the angle on a surface where liquid tends to invade cavities (as depicted in Figure 7). Then we have $\phi_1 = \phi_S$, $\theta_1 = \theta$, $\phi_2 = 1 - \phi_S$, and $\theta_2 = 0$ (1 being the solid, and 2 the liquid), so that we get:

$$\cos\theta^* = 1 - \phi_S + \phi_S \cos\theta. \tag{2.10}$$

This behavior is in good qualitative agreement with the observations reported in Figure 6. When the contact angle becomes smaller than some quantity, the Wenzel behavior is replaced by another regime, where $\cos\theta^*$ is found to increase linearly with $\cos\theta$, with a slope smaller than 1, and joins the value $\theta^* = 0$ only for $\theta = 0$. This suggests that the value of ϕ_S, of obvious practical interest (since it represents the fraction of solid in contact with the liquid), which is difficult to predict on a disordered surface such as the one used in the experiments of Figure 6, can be easily measured. However, this is

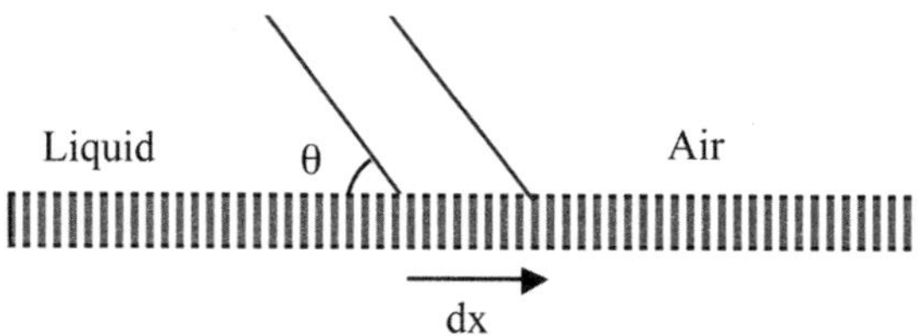

Figure 8. Edge of a drop deposited on a solid decorated by chemical stripes, a grey and a white one, characterized by a surface fraction ϕ_1 and ϕ_2 ($\phi_1 + \phi_2 = 1$) and contact angle θ_1 and θ_2, respectively.

only true if equation (2.10) is satisfied, i.e. if the energy of the system can find its ground state. As for the Wenzel situation (equation (2.7)), this is far from being obvious: as stressed by Marmur, these different situations have multiple states of equilibrium (which is physically due to the possibility for the contact line to pin on the roughness or chemical discontinuity), making it possible that these 'ideal' models are not satisfied (Marmur, 2006).

The ideas are quite similar in the hydrophobic case. There again, two states are likely to exist: a Wenzel state, given by equation (2.7), or a Cassie-Baxter state, sometimes referred to as the 'fakir state', because it corresponds to the situation of a drop sitting on the little posts (with air below) as the fakir does on the bed of nails. Then, the contact angle should be given by equation (2.8), taking $\phi_1 = \phi_S$, $\theta_1 = \theta$, $\phi_2 = 1 - \phi_S$, and $\theta_2 = \pi$ (1 is still the solid, but 2 is air), which yields:

$$\cos\theta^* = -1 + \phi_S + \phi_S \cos\theta. \tag{2.11}$$

This angle will be very large, of the order of 160° to 170°, provided that ϕ_S is small (around 10%). This is indeed the case, as observed in Figure 9 where a drop of water is deposited on a solid consisting of thin dilute posts ($\phi_S = 2\%$), the whole substrate treated to be hydrophobic. Using backlighting, and owing to the high dilution of pillars, it is even directly seen that the drop sits at the top of the posts, as assumed in our description.

As done in the hydrophilic domain, a comparison between the respective energies of the two states indicates which state should be selected by the system: a Wenzel state for moderate hydrophobicity or roughness (then, it is not too costly energetically for the liquid to follow the surface of the solid), or a fakir state in the opposite limit. However, as shown in Figure 6 where a discontinuity of the contact angle is observed as the hydrophobic domain is entered, fakir drops can exist even in regions where the

Figure 9. Millimetric water drop deposited on a solid decorated with micrometric hydrophobic pillars of height $h = 25$ μm, and dilution $\phi_S = 0.02$). The contact angle is found to be very high (of the order of 170°), and the distance between the drop and its reflection measured to be 50 μm, showing that the drop indeed sits on the top of the pillars (Photo by Mathilde Callies).

Wenzel state should be the most stable - indicating that this state can be metastable (this is the case for example in Figure 9). Then comes naturally the question of the robustness of this metastability: energy barriers have to be overcome to find the ground state, and different tests were recently proposed to quantify this robustness (Callies and Quéré, 2005).

These questions might be quite important practically. Both superhydrophobic states (that is, Wenzel and fakir) have a very different property. The fakir state is characterized by a very small contact angle hysteresis (because the liquid only interacts with the top of the posts, whose proportion will often be small), often of the order of 5 to 20°, while the hysteresis in the Wenzel state can be of the order of 100° or even more. The consequences are obvious: a fakir drop will easily run down inclines, and will most often rebound at impact, which defines water repellency. On the other hand, a Wenzel drop will remain stuck on inclines (despite its large advancing angle), and it will get pinned at impact, providing a very imperfect rebound. This difference between a sticky and a slippery state should also concern water flowing on such substrate. A first series of experiments showed that a significant slippage (with so-called 'slip lengths' of a few tens of micrometers, instead of a few nanometers on smooth solids) might be observed on these textured solids, provided that a fakir state is achieved (Ou et al., 2004). In the opposite case, no slippage is expected (Cottin-Bizonne et al., 2003).

3 Dynamics

We consider in this section three different canonical problems, related to the dynamics of wetting. We first describe the simple case of the dynamics of wicking, and discuss recent extensions. Then, we briefly comment on the dynamics of wetting on flat surfaces, and finally present the classical laws of coating.

3.1 Wicking

One first simple and important interfacial flow concerns wicking - that is, the behavior of a wetting liquid contacting a porous medium, and penetrating it. We all observed such phenomena, for example as oil contacts a fabric (our shirt): the wicking is first very quick, and then slows down, independently of gravity; on our shirt, the oily spot remains circular as it expands, showing that gravity is negligible, as expected at scales smaller than the height of capillary rise, which is very large for such small pores.

The first model for this process was proposed in the twenties by Washburn (1921). The porous medium is viewed as a collection of parallel little tubes (of radius R), which all drive the liquid at the same rate. For a wetting liquid, the driving force can be seen as a Laplace pressure $2\gamma/R$ integrated on the surface area πR^2, which yields a force $2\pi R\gamma$. Balancing this force with the classical Poiseuille formula $8\pi\eta V z$ for the viscous resistance (with η the liquid viscosity, V the velocity of the front and z its position in the tube), we find that the meniscus progresses with a diffusive-type law:

$$z = \left(\frac{\gamma R t}{2\eta}\right)^{\frac{1}{2}}. \tag{3.1}$$

This behavior roughly agrees with our observations on real porous media (fabric, paper, sand, etc.), despite the complexity of these media. We could naively assume that a microscopic observation of the medium provides a mean 'tube' size, equation (3.1), but it is generally far from being that simple. For example, if we consider the case of bidisperse tubes connected with each other (which might be relevant for fabrics consisting of porous yarns woven together), we find that two fronts propagate: a main front (as for the mass), corresponding to the filling of the big tubes, and following equation (3.1) (with the radius R_1 of these tubes), which follows another front, ahead of it, which also progresses as the square root of time and corresponds to the propagation of liquid in the small tubes (of radius R_2) which use the big ones as reservoirs. The distance between the two fronts is also given by an equation such as equation (3.1), but its characteristic length is R_2^2/R_1, much smaller than the two natural sizes (namely, R_2 and R_1) of the system (Bico and Quéré, 2003).

Another natural question concerns the progression of a film on a texture, such as sketched in Figure 7. There again, it is observed that the front position increases as the square root of time, but we expect that both the driving and the resisting forces should depend on the pattern of the surface. We emphasized for example that the energy change associated with the progression of the film by a quantity dx writes, $dE = (\gamma_{SL} - \gamma_{SA})(r - \phi_s)dx + \gamma(1 - \phi_s)dx$, where r is the solid roughness and ϕ_s the proportion of solid which remains. A force can be easily deduced from this expression, and this force (written as the energy per unit length) simplifies to $F = (r - 1)\gamma$, in a complete wetting situation ($\phi_s = 0$ and $\theta = 0$). This expression indeed reflects the presence of a texture, through the parameter r. For dilute pillars (of mutual distance larger than their height), the viscous resistance is set by the pillar height (that is, the thickness of the flow), which makes this force scale as $\eta V x / h$, per unit length. The balance of these two forces thus yields a diffusive-type law, whose coefficient varies as a function of the nature of the pattern (C. Ishino et al., 2006).

3.2 A Few (Moving) Drops

The dynamics of drops has been a very active field for the last 20 years (de Gennes et al., 2004). Both the spreading of drops in complete wetting, and the motion of drops in partial wetting, yet subjected to a force, were considered. Our aim here is not to review these discussions, but rather to give simple arguments to obtain orders of magnitude for these moving globules.

As early emphasized by Huh and Scriven (1971), the viscous force which resists the motion of a drop primarily comes from the presence of moving contact lines, close to which the liquid thickness vanishes, thus making the viscous force likely to diverge. We shall consider here only the limit of small velocities (often observed experimentally), for which the contact angle remains close to its static value θ (we are in partial wetting) – see Figure 10. Then, the viscous force writes as the liquid viscosity times the velocity gradient, which must be integrated over the surface area of the drop. Considering the vicinity of the contact line as a wedge, of equation $z = \theta x$, we deduce a viscous force (per unit length) scaling as $\int dx \eta V / \theta x$. This expression makes it clear that the viscous force diverges at the contact line (the integration must be done between $x = 0$ (position of the line) and $x = R$ (typical size of the drop)). However, it also allows to understand that this divergence is weak, since it is logarithmic. Provided that a cut-off (of the order of a molecular size) is introduced, this logarithmic term should be of the order of 10 to 15, and treated as a constant, owing to the very slow variation of this function (de Gennes et al., 2004). We denote this number as α.

We can now obtain the velocity of the moving drop. If it is driven by gravity (drop of size R deposited on a plate tilted by a small angle β), the driving force written per unit length scales as $\rho g R^2 \beta$, so that the velocity is given by:

$$V \sim \frac{\rho g R^2 \beta}{\eta \theta \alpha}. \tag{3.2}$$

This expression is even simpler if the drop is driven by a capillary force (of the order of γ, per unit length), as in the cases considered in Section 2.2. Then we find:

$$V \sim \frac{\gamma}{\eta \theta \alpha}. \tag{3.3}$$

This formula is remarkable since it does not depend on the drop size (which fixes both the viscous and the capillary forces). All these considerations (equations (3.2) and (3.3))

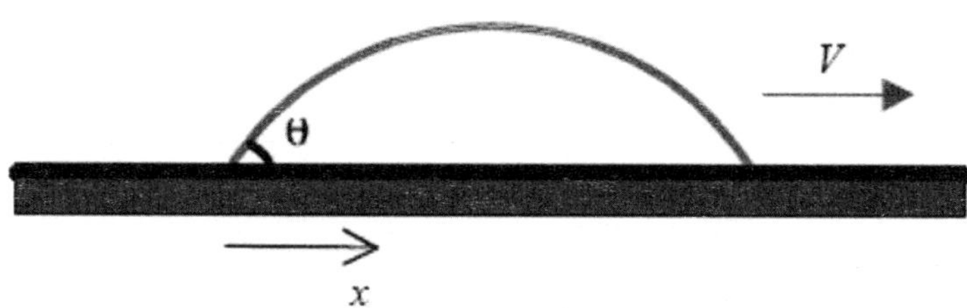

Figure 10. A small drop moving in a chemical or gravitational field. The velocity is small enough to assume that the drop meets the solid with its static angle θ.

should remain valid provided that the capillary number remains smaller than unity. If this condition is not satisfied, the drop becomes asymmetric, with an angle larger at the front than at the rear, meaning that the angles are a function of the velocity, which of course complicates the discussion. More generally, the question of the dynamic angle remains one of the open fields of interfacial hydrodynamics, and many strong debates still divide the community.

3.3 Fluid Coating

The key experiment consists of drawing a solid out of a bath of wetting liquid. Then the solid comes out coated by a liquid film, whose thickness can be adjusted by playing with the withdrawal velocity. We sketch in Figure 11 this coating device.

Goucher and Ward (1922) identified the key parameters of the problem in a seminal series of experiments performed with a fiber as a solid. On the one hand, the boundary condition at the *solid-liquid interface* is responsible for the liquid entrainment: owing to the liquid viscosity, the liquid close to the solid must move at the same velocity as the solid (and thus comes out with it). On the other hand, the motion of the solid provokes a deformation of the *liquid-air interface* (by a length ℓ in the figure), that surface tension opposes. Thus, viscous and capillary forces play antagonist roles, and it is natural to consider as a parameter the ratio of these forces, the so-called *capillary number*, which writes:

$$\mathrm{Ca} = \frac{\eta V}{\gamma} \tag{3.4}$$

where η and γ are the liquid viscosity and surface tension, and V the coating velocity. Thus, it is expected that the film thickness writes:

$$h = af(\mathrm{Ca}) \tag{3.5}$$

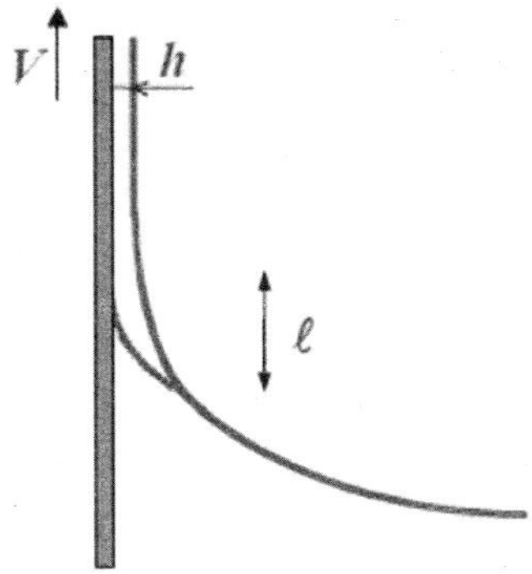

Figure 11. Withdrawing a plate of a bath of wetting liquid produces a coating. In the limit of small coating velocities, only the top of the static meniscus is affected by the deposition: this is the limit considered by Landau and Levich (1942) for calculating the film thickness.

where a is some static length. This length can only be the capillary length, which sets the size of the meniscus from which the solid is extracted. Goucher and Ward obtained data by coating metallic wires with melt beeswax and weighing them after solidification, and found that $f(\text{Ca})$ increases with Ca (Goucher and Ward, 1922).

The first (correct) determination of the function $f(\text{Ca})$ in equation (3.5) is the theory developed by Landau, Levich and Derjaguin (LLD theory) (Landau and Levich, 1942; Derjaguin, 1943). We summarize it by using scaling arguments. The description is based on Figure 11. At a small velocity, the static meniscus is hardly deformed by the entrained film. In other words, the region matching this meniscus with the film has an extension ℓ much smaller than the meniscus height κ^{-1} (first LLD hypothesis). This region is curved, which induces a Laplace depression opposing the coating. The second LLD hypothesis consists in assuming that the associated pressure gradient, of the order of $\gamma h/\ell^3$, is larger than the gravity force ρg, which also opposes the liquid entrainment. In the thin film approximation ($h \ll \ell$, third LLD hypothesis), and assuming a small Reynolds number (fourth LLD hypothesis), the Navier-Stokes equation can be thus written dimensionally:

$$\frac{\eta V}{h^2} \sim \frac{\gamma h}{\ell^3}. \tag{3.6}$$

The dynamic region where the film is formed matches the static meniscus close to its top (first LLD hypothesis). The balance of the pressures at the matching point thus writes:

$$\rho g \kappa^{-1} \sim \frac{\gamma h}{\ell^2}. \tag{3.7}$$

We deduce from the latter equation that ℓ scales as $(h\kappa^{-1})^{1/2}$. Introducing this expression in equation (3.6) allows us to obtain the Landau-Levich law (Landau and Levich, 1942):

$$h = 0.94\kappa^{-1}\mathrm{Ca}^{2/3} \qquad (3.8)$$

where the numerical coefficient calculated by Derjaguin (1943) was introduced, together with the capillary number (equation (3.4)).

Putting together equations (3.7) and (3.8), we find $\ell \sim \kappa^{-1}\mathrm{Ca}^{1/3}$. This allows us to check the validity of the different hypotheses. Remarkably, it turns out that the three first LLD hypothesis are the same one, writing $\mathrm{Ca}^{1/3} \ll 1$, that is, Ca smaller than 10^{-3}. For many liquids, this condition is not so restrictive. For water for example, it imposes withdrawal velocities smaller than 10 cm/s. If it is not satisfied, the weight cannot be neglected as a force opposing the liquid entrainment. As first pointed out by Derjaguin (1943), the film thickness then simply results from a balance between viscous force $(\eta V/h^2)$ and gravity (ρg), which immediately yields:

$$h \sim \kappa^{-1}\mathrm{Ca}^{1/2} \qquad (3.9)$$

which should be valid in the opposite limit $(\mathrm{Ca}^{1/3} \gg 1)$, with a long transition regime between the two asymptotic formulae (3.8) and (3.9).

Note finally that we also assumed a small Reynolds number. This condition also implies small withdrawal velocities: inertial terms in this problem are of the order of $\rho V^2/\ell$, while viscous ones (as expressed in equation (3.6)) scale as $\eta V/h^2$. Using the LLD scaling for h and ℓ, we thus find that a small Reynolds number corresponds to $\rho V^2 \kappa^{-1}/\gamma \ll 1$, a condition independent of the liquid viscosity! For most liquids, the velocity $(\gamma/\rho\kappa^{-1})^{1/2}$ is nearly constant, and found to be of the order of 10 cm/s.

We show in Figure 12 the results of a series of reflectometry measurements, which give the thickness of a film of oil entrained by smooth plates withdrawn out of a bath of silicone oil (M. Callies et al., 2006). Using the capillary number as a control parameter (both the oil viscosity and the plate velocity were varied in the experiment), we find that the Landau-Levich law (drawn with a line) is remarkably well obeyed in the range where it is expected to be valid $(\mathrm{Ca} < 10^{-3})$. For larger capillary numbers, the film is found to be slightly thinner, as expected from the action of gravity.

The Landau-Levich law is ubiquitous, and there are many situations where it is observed with some variations. For example, liquids moved in tubes similarly deposit a film behind, and the thickness of this film is also given by a Landau-Levich argument, provided that we take the tube radius as the characteristic length normalizing the thickness (Bretherton, 1961). In a similar spirit, the thickness of a freshly made soap film should obey the Landau-Levich scaling, if the surfactant layers play the role of a wall, as postulated by Frankel and co-workers (and often verified in experiments) (Mysels et al., 1959). But one rather unexpected situation, which also leads to such a law, is the case of air entrainment.

We have all observed that diving in a swimming-pool produces air bubbles, while entering slowly in the same pool does not: the motion of a solid penetrating a liquid at a high speed induces air entrainment. This phenomenon often defines the maximum speed of coating of a solid, which has to enter a bath before leaving it coated. In general, air

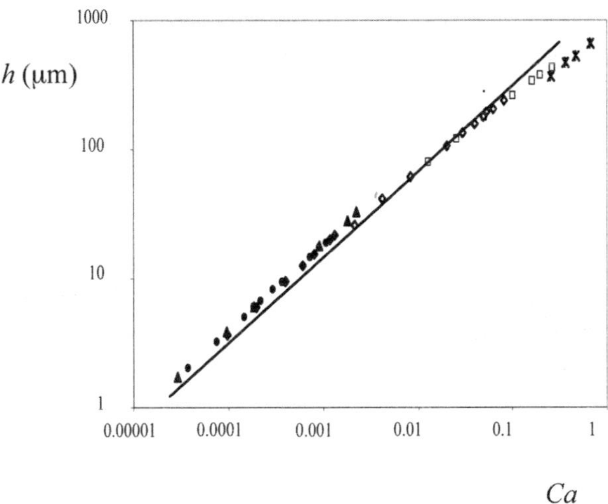

Figure 12. Measured thickness of a film of silicone oil entrained by a silicon wafer withdrawn out of a reservoir. Many different oils are used, and all the data collapse in the same curve provided they are plotted as a function of the capillary number, $Ca = \eta V / \gamma$. The data obey the Landau-Levich law (equation (3.8), drawn in full line) at small Ca, but deviate as the capillary number increases.

entrainment must be avoided because of the resulting bubbles (which are all the more harmful since the coating solution generally contains surfactants, and thus is likely to transform to a foam). The same phenomenology applies for a jet impinging a bath of the same nature. For example, if we fill a glass of water from a small or large height, we observe that air is injected with the jet only for the large height - there again, there is a threshold velocity above which air is entrained.

A natural question is the rate of air entrainment. We shall only consider the limit of viscous liquid. Then, it is particularly relevant to compare photos showing the deformation of the interface (just) below, and (just) above the threshold of entrainment (Figure 13). Just below the entrainment, it is observed that the jet largely deforms the surface, on a depth of about 1 cm. However, the situation is dramatically different once air is entrained (Lorenceau et al., 2004): the jet is found to penetrate the bath coated with a film, down to a depth of a few centimeters, but the main fact is the 'relaxation' of the outside meniscus: the film of air decouples the flow (i.e. the jet) from the rest of the bath, which behaves as if it were contacting a non-wetting solid (because a liquid does not wet air, as discussed in the section 2.3). Hence, this is an inverted Landau-Levich situation (upside down), where the solid-like jet entrains a thin layer of air in a 'passive' viscous 'atmosphere', that is, the rest of the bath. We thus expect a Landau-Levich film of air, with the difference that it is made in a viscous medium rather than in an inviscid one (as assumed in Figure 12). This modifies the boundary condition at the liquid/vapour interface, but not the physics of the process, which eventually provides a Landau-Levich scaling yet a different numerical coefficient - as indeed observed experimentally.

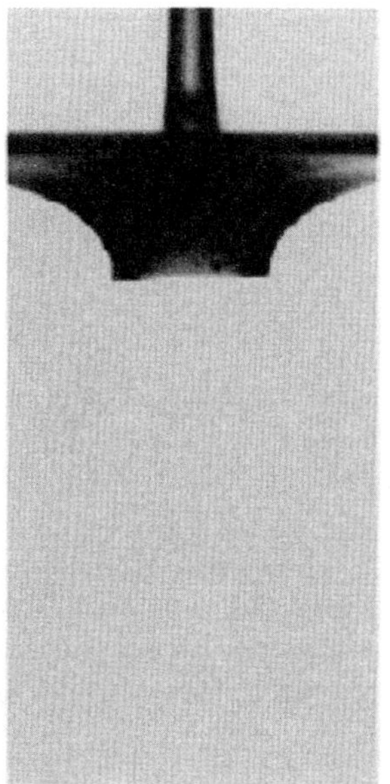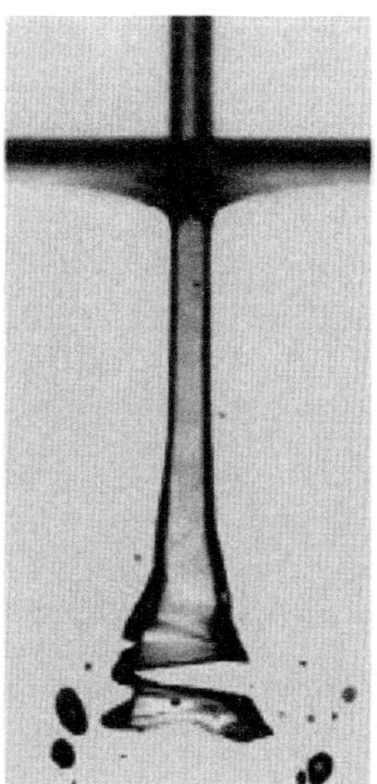

Figure 13. Jet of silicone oil (1000 more viscous than water) hitting a bath of the same nature. The photos show a side view of the experiment just below the threshold of air entrainment (left), and just above (right) (Photos: Elise Lorenceau).

4 Conclusion

As most often in Soft Matter Physics, problems frequently originate from practical issues. Water-repellency, coating devices, air entrainment are such questions, and we showed here how simple answers can be proposed to these complicated questions. These views rarely solve industrial issues (often related to a very particular device), but they often show which are the relevant parameters, and sometimes help people in companies to think differently. Conversely, these kinds of questions are a source of inspiration for researchers, and their generality and importance explains why these problems have been treated periodically (hopefully with some progress along the decades) for the last 100 years. The development of a new applied science, such as microfluidics, at the scale of which surface forces are most often dominant, has also contributed to irrigate this field. On the whole, despite its old age, three-phase capillarity should still have a flourishing future.

Acknowledgements

I thank S. Kalliadasis and U. Thiele for their active participation and very kind patience, M. G. Velarde for his encouragement, M. Callies and E. Lorenceau for many discussions and precious help.

Bibliography

C. D. Bain, G. D. Burnett-Hall, and R. R. Montgomerie. Rapid motion of liquid drops. *Nature*, 372:414–414, 1994.

J. Bico and D. Quéré. Precursors of impregnation. *Europhys. Lett.*, 61:348–353, 2003.

J. Bico, U. Thiele, and D Quéré. Wetting of textured surfaces. *Colloid Surf. A*, 206: 41–46, 2002.

F. P. Bretherton. The motion of long bubbles in tubes. *J. Fluid Mech.*, 10:166–188, 1961.

M. Callies and D. Quéré. On water repellency. *Soft Matter*, 1:55–61, 2005.

A. B. D. Cassie and S. Baxter. Wettability of porous surfaces. *Trans. Faraday Soc.*, 40: 0546–0550, 1944.

C. Cottin-Bizonne, J. L. Barrat, L. Bocquet, and E. Charlaix. Low-friction flows of liquid at nanopatterned interfaces. *Nat. Mater.*, 2:237–240, 2003.

S. Daniel and M. K. Chaudhury. Rectified motion of liquid drops on gradient surfaces induced by vibration. *Langmuir*, 18:3404–3407, 2002.

S. Daniel, M. K. Chaudhury, and P. G. de Gennes. Vibration-actuated drop motion on surfaces for batch microfluidic processes. *Langmuir*, 21:4240–4248, 2005.

P.G. de Gennes, F. Brochard-Wyart, and D. Quéré. *Drops, Bubbles, Pearls and Waves*. Springer, New York, 2004.

B.V. Derjaguin. On the thickness of the liquid film adhering to the walls of a vessel after emptying. *Acta Physicochim. URSS*, 20:349–352, 1943.

F. Domingues Dos Santos and T. Ondarçuhu. Free-running droplets. *Phys. Rev. Lett.*, 75:2972–2975, 1995.

F. S. Goucher and H. Ward. A problem in viscosity - the thickness of liquid films formed on solid surfaces under dynamic conditions. *Philos. Mag.*, 44:1002–1014, 1922.

C. Huh and L. E. Scriven. Hydrodynamic model of steady movement of a solid / liquid / fluid contact line. *J. Colloid Interface Sci.*, 35:85–101, 1971.

M. Callies, M. Maleki, F. Restagno and D. Quéré. *to be published*, (2006).

L. Landau and B. Levich. Dragging of a liquid by a moving plate. *Acta Physicochim. URSS*, 17:42–54, 1942.

E. Lorenceau, D. Quéré, and J. Eggers. Air entrainment by a viscous jet plunging into a bath. *Phys. Rev. Lett.*, 93:254501, 2004.

C. Ishino, M. Callies, K. Okumura, and D. Quéré. *to be published*, (2006).

A. Marmur. Soft contact: measurement and interpretation of contact angles. *Soft Matter*, 2:12–17, 2006.

K.J. Mysels, K. Shinoda, and S. Frankel. *Soap Films*. Pergamon Press, London, 1959.

J. Ou, B. Perot, and J. P. Rothstein. Laminar drag reduction in microchannels using ultrahydrophobic surfaces. *Phys. Fluids*, 16:4635–4643, 2004.

J. Rowlinson and B. Widom. *Molecular theory of capillarity*. Oxford University Press, Oxford, 1982.

S. Shibuichi, T. Onda, N. Satoh, and K. Tsujii. Super water-repellent surfaces resulting from fractal structure. *J. Phys. Chem.*, 100:19512–19517, 1996.

Y. Sumino, N. Magome, T. Hamada, and K. Yoshikawa. Self-running droplet: Emergence of regular motion from nonequilibrium noise. *Phys. Rev. Lett.*, 94(6):068301, 2005.

E. W. Washburn. The dynamics of capillary flow. *Phys. Rev.*, 17:273–283, 1921.

R. N. Wenzel. Resistance of solid surfaces to wetting by water. *Ind. Eng. Chem.*, 28: 988–994, 1936.

T. Young. An essay on the cohesion of fluids. *Phil. Trans. R. Soc.*, 95:65–87, 1805.

Falling Films Under Complicated Conditions

Serafim Kalliadasis

Department of Chemical Engineering, Imperial College London, London, United Kingdom

Abstract A film falling down an inclined plane has been an active topic of fundamental research at the international level for several decades both theoretically and experimentally and is now a classical hydrodynamic instability problem. However, the dynamics of a falling film in the presence of additional complexities, as compared to the classical problem, has largely been ignored by the majority of studies in falling films, even though these complexities are crucial in most situations of practical interest. These additional factors include heated substrates and three-dimensional effects. Here we present very recent and most-up-to-date developments for the problem of a falling film in the presence of these complexities and we outline open questions and issues which have not been resolved.

1 Film Falling Down a Heated Plate

1.1 Introduction

A film falling down a planar substrate exhibits a rich variety of spatial and temporal structures. It is a convectively unstable open-flow hydrodynamic system with a sequence of wave transitions that begins with amplification of small-amplitude white noise at the inlet, filtering of linear stability, secondary modulation instability that transforms the primary wave field into a solitary pulse and inelastic pulse-pulse interaction. This evolution is driven by the classical long-wave instability mode first observed in the pioneering experiments by Kapitza and Kapitza (1949). This mode was analyzed in detail by Benjamin (1957) who determined its threshold and showed that a film falling down a plane inclined at an angle β with respect to the horizontal direction, can only be destabilized for a Reynolds number larger than the critical value $5/6 \cot \beta$ (with the Reynolds number based on the flow rate).

Benney (1966) was the first to apply the long-wave expansion (LWE) to this problem to obtain a single equation of the evolution type for the free surface (see also the discussion by Thiele in this book). In his expansion he took the Weber number to be of $O(1)$ with respect to the long-wave parameter ϵ. The correct second order LWE was obtained by Gjevik (1971) and Lin (1974) who both took the Weber number to be of $O(\epsilon^{-2})$. Later on Pumir et al. (1983) and Nakaya (1989) constructed numerically solitary wave solutions of the first order LWE and demonstrated that the solitary wave solution branches for the speed of the waves as a function of Reynolds number show branch multiplicity and limit points above which solitary waves do not exist. Time-dependent computations by Pumir et al. (1983) demonstrated that LWE exhibits finite-time blow-up behavior for a sufficiently large set of smooth initial data when this equation is integrated in regions of the parameter space where solitary waves do not exist. Obviously, this behavior is unrealistic and marks the failure of LWE to correctly describe nonlinear waves far from criticality – close to criticality LWE is exact; this is not surprising as LWE is a regular perturbation expansion of the full Navier-Stokes equations. The connection between the absence of solitary wave solutions and finite-time blow up was demonstrated clearly by Rosenau et al. (1992) and Oron and Gottlieb (2002) and further investigated by Scheid et al. (2005b). Ooshida (1999) on the other hand was able to regularize LWE but the resulting wavespeeds in the region of moderate Reynolds numbers were much smaller than those obtained from full Navier-Stokes.

Following the pioneering theoretical work by Kapitza (1948), an ad-hoc but convenient simplification was employed by Shkadov (1967, 1968) who developed the integral-boundary-layer (IBL) approximation which combines the boundary-layer approximation of the Navier-Stokes equations assuming a self-similar parabolic velocity profile and long waves on the interface (see e.g. Chang and Demekhin (2002)) with the Kármán-Pohlhausen averaging method in boundary-layer theory. This procedure results in a two-equation model for the free surface and local flow rate and unlike LWE, IBL has no limit points and predicts the existence of solitary waves for all Reynolds numbers. In fact the IBL solitary wave solution branches are in quantitative agreement with the boundary-layer (Demekhin et al., 1987) and full Navier-Stokes equations (Demekhin and Kaplan, 1989; Salamon et al., 1993; Ramaswamy et al., 1996). A detailed review of wave formation on a free falling film is given by Chang (1994) and Chang and Demekhin (1996, 2002).

However, despite its success to describe nonlinear waves far from criticality, Shkadov's IBL approach does have some shortcomings with the principal one being an erroneous prediction of the critical Reynolds number. By combining a gradient expansion with a weighted residual technique using polynomial test functions, Ruyer-Quil and Manneville (2000, 2002) obtained a two-equation model having the same structural form as Shkadov's but recovering correctly the instability threshold. The same authors also developed higher order models taking into account the second-order viscous dissipative effects.

The onset of the instability for a film falling down a uniformly heated wall, was analyzed in detail by Goussis and Kelly (1991). In this case the wall heating generates a temperature distribution on the free surface which in turn induces surface tension gradients that affect the free surface and therefore the fluid flow. Goussis and Kelly performed a linear stability analysis based on Orr-Sommerfeld and linearized energy equations. They provided a detailed numerical solution of the pertinent eigenvalue problem and demonstrated that in addition to the Kapitza hydrodynamic mode of instability, the heated falling film is also subject to two thermocapillary instability modes: a short-wave mode obtained first by Scriven and Sternling (1964) who considered the thermocapillary instability of an horizontal layer with a non-deformable free surface and a long-wave mode first obtained by Pearson (1958) who allowed the free surface to deform.

The nonlinear stage of the instability for the uniformly heated falling film problem was investigated by Joo et al. (1991) who utilized the LWE to obtain an evolution equation for the film thickness. In addition to the Marangoni effect, they also included evaporation effects and long-range attractive intermolecular interactions. These authors also performed time-dependent computations which indicate finite-time blow up behavior – referred to as 'super-exponential' or 'catastrophic' by the authors – and is obviously related to the failure of LWE. A review of a wide variety of fluid flow problems using LWE including problems with Marangoni effects is given by Oron ct al. (1997) (see also the chapter by Thiele in this book).

A detailed investigation of the dynamics of a film falling down a uniformly heated wall was recently undertaken by Kalliadasis et al. (2003a). Their analysis was based on the model equations derived by Kalliadasis et al. (2003b) in their study of the thermocapillary instability of a thin liquid film heated from below by a local heat source. These authors formulated an IBL approximation of the equations of motion and energy equation by adopting a linear test function for the temperature field combined with a weighted residuals approach for the energy equation to obtain a three-equation model for the free surface, local flow rate and interfacial temperature. However, despite the success of this IBL model in the nonlinear regime, it does not predict very accurately neutral and critical conditions and hence it suffers from the same limitations with the Shkadov model for the isothermal film. The limitations of the model equations derived by Kalliadasis et al. (2003a,b) were recently overcome by Ruyer-Quil et al. (2005) and Scheid et al. (2005a). In addition, these authors also took into account the second-order dissipative effects both in the momentum and energy equations. These second-order terms were neglected in the formulation by Kalliadasis et al. (2003a,b) while they indeed play an important role in the dispersion of waves for larger Reynolds numbers. The procedure followed is effectively an extension of the methodology applied in the case of isothermal flows by Ruyer-Quil and Manneville (2000, 2002) and is based on a high-order weighted residuals approach with polynomial expansions for both velocity and temperature fields. Details of the theoretical developments are also given in the thesis by Scheid (2004).

Here we revisit the heated falling film problem. We use the thesis by Scheid (2004) as a starting point. We impose two types of wall boundary conditions: a heat flux (HF) and a specified temperature (ST) condition. Note that all previous studies on the heated falling film problem imposed the ST condition only. Scheid's PhD thesis is the first study that introduced HF. We employ the same first order in ϵ single-mode Galerkin representation for the transport of momentum given by equations (5.17a) and (5.17b) in Scheid (2004) (and equations (4.18a) and (4.18b) in Ruyer-Quil et al. (2005)). However, for the transport of heat we develop a refined treatment of the energy equation that results in an alternative system of first order in ϵ energy equations obtained by introducing test functions which now satisfy all boundary conditions so that our Galerkin approach incorporates all boundary conditions within its projection. On the other hand, the models derived in Scheid (2004) and Ruyer-Quil et al. (2005) adopted test functions which do not satisfy all boundary conditions. The Galerkin projection then incorporated the boundary conditions on the boundary terms resulting through integrations by parts following the averaging of the energy equation. Moreover, unlike the studies by Scheid (2004) and Ruyer-Quil et al. (2005) where the amplitudes in the expansion for the temperature are assigned certain orders with respect to ϵ, in our projection for the temperature the order of the amplitudes is not specified.

We demonstrate that the linear stability properties of a three-equation model for the free surface, local flow rate and interfacial temperature (obtained from a single-mode Galerkin projection of the energy equation) are in good agreement with an Orr-Sommerfeld analysis of the linearized Navier-Stokes and energy equations. We also develop an LWE model (a single equation of the evolution type for the free surface) which is used to ensure that the models obtained from our weighted residuals approach predict the correct behavior close to criticality. We demonstrate that by taking a sufficient number of amplitude equations we can obtain with an appropriate gradient expansion LWE, but increasing the number of amplitude equations significantly complicates our system. Nevertheless, a comparison with the numerical solution of the full energy equation at first order in ϵ shows that the single-mode Galerkin projection provides a substantially improved representation of the temperature field (and hence the flow field due to the coupling through the Marangoni effect) over the first-order temperature models given by equation (8.8) for HF and equation (5.17c) for ST from Scheid (2004) (see also equations (4.18a-c) for ST in Ruyer-Quil et al. (2005)).

We also construct bifurcation diagrams for permanent solitary waves and demonstrate that unlike LWE that exhibits limit points and branch multiplicity, our three-equation model and the three-equation model given in Scheid (2004) and Ruyer-Quil et al. (2005) predict the continuing existence of solitary waves for all Reynolds numbers (albeit with slightly different solitary wave speeds between the different weighted residual models at higher Reynolds numbers). The good agreement between our model and the numerical solution of the full energy equation at first order in ϵ persists up to a certain Reynolds number at which a recirculation zone appears in the crest of a solitary wave. Examination of the numerical solution of the energy equation indicates that further increase of the Reynolds number leads to the formation of a sharp peak for the interfacial temperature associated with the creation of a thermal boundary layer in the vicinity of the stagnation point at the front of a solitary wave. As we further increase the Reynolds number we observe an increased deviation between the interfacial temperature predicted by our model and the numerical solution of the energy equation. Indeed, to accurately represent the temperature field as the boundary layer develops, one would need an increasingly large number of test functions. At some point and as the thickness of the boundary layer tends to zero the number of test functions should tend to infinity. Hence, any weighted residual approach is bound to fail in this region. However, we do expect that physically it will be rather difficult for the system to sustain a two-dimensional thermal boundary layer and it is quite likely that three-dimensional effects and related three-dimensional instabilities (e.g. rivulet formation) will diffuse the sharp temperature peaks in the transverse direction.

1.2 Problem Definition, Scalings and Governing Equations

We consider the dynamics of a thin liquid film falling down a planar heated wall, as illustrated in Figure 1. The wall forms an angle β with the horizontal direction. The heating is provided by a heat source inside the plate that uniformly generates heat, e.g. an electric heating device. The surrounding gas phase below the plate and above the liquid film is air maintained at the constant far field temperature T_a. For simplicity we shall employ approximate boundary conditions at the liquid-gas and liquid-wall interfaces, thus bypassing the much more involved conjugated heat transfer problems in the air and

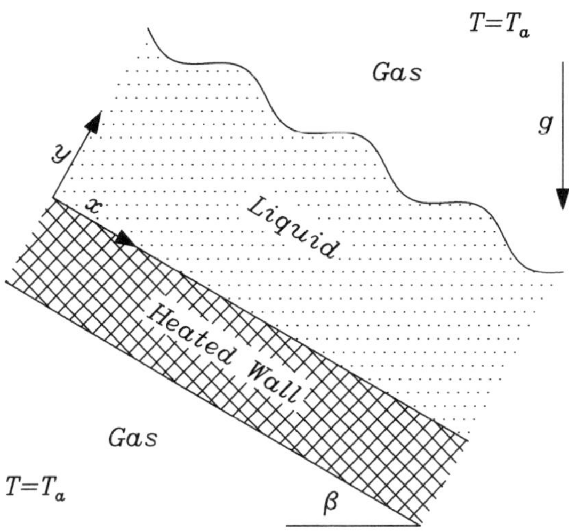

Figure 1. Sketch of the profile geometry for a thin liquid film falling down an inclined heated wall forming an angle β with the horizontal direction. The surrounding gas phase is maintained at temperature T_a.

the plate: on the free surface we assume Newton's law of cooling while on the wall we shall focus on two types of boundary conditions, HF and ST, as discussed in the previous section. For HF the heater is assumed to generate a constant heat flux q_0, while for ST the heater is assumed to maintain the wall temperature at the constant value T_w. The heating provided by the wall induces a thermocapillary Marangoni effect at the interface which affects the interface and therefore the fluid flow.

The liquid has viscosity μ, density ρ, constant pressure heat capacity c_p, thermal diffusivity κ and thermal conductivity $\lambda = \rho c_p \kappa$, all assumed to be constant. We also assume that the liquid is non-volatile so that evaporation effects can be neglected while the film is sufficiently thin so that buoyancy effects can be neglected. The governing equations are conservation of mass, Navier-Stokes and energy equations,

$$\underline{\nabla} \cdot \underline{u} = 0, \quad \underline{u}_t + (\underline{u} \cdot \underline{\nabla})\underline{u} = -\frac{1}{\rho}\nabla p + \nu \nabla^2 \underline{u} + \underline{g} \qquad (1.1a, b)$$

$$T_t + (\underline{u} \cdot \underline{\nabla})T = \kappa \nabla^2 T, \qquad (1.1c)$$

where $\underline{u}$, p and T are the velocity, pressure and temperature of the liquid, respectively. $\underline{g}$ is the gravitational acceleration and $\nu = \mu/\rho$ is the kinematic viscosity of the liquid. A coordinate system (x, y) is chosen so that x is the streamwise coordinate and y is the outward-pointing coordinate normal to the wall. The wall is then located at $y = 0$ and the interface at $y = h(x, t)$.

On the wall, we have the usual no-slip and no-penetration boundary conditions

$$\underline{u} = 0 \quad \text{on} \quad y = 0 \qquad (1.2a)$$

while the boundary condition for the temperature field depends on the type of problem being examined. For the heat flux problem the thermal boundary condition is (Scheid, 2004)

$$\text{HF:}\ \lambda\underline{\nabla}T \cdot \underline{j} = -q_0 + \alpha_w(T - T_a) \quad \text{on}\quad y = 0 \tag{1.2b}$$

where $\underline{j}$ is the unit vector normal to the wall and pointing into the liquid and α_w is the heat transfer coefficient between the wall and the air below the wall, assuming that the wall has effectively zero thickness. This mixed boundary condition implies that both the flux supplied by the solid substrate to the liquid q_0 and the heat losses $\alpha_w(T - T_a)$ to the ambient gas phase contribute to the temperature gradient at $y = 0$. On the other hand, for the specified wall temperature problem the thermal boundary condition is simply

$$\text{ST:}\ T = T_w \quad \text{on}\quad y = 0. \tag{1.2c}$$

On the interface we have the kinematic boundary condition along with the normal and tangential stress balances

$$h_t + \underline{u} \cdot \underline{\nabla}(h - y) = 0, \quad p_a + \underline{\underline{\tau}} \cdot \underline{n} \cdot \underline{n} = -\sigma \underline{\nabla} \cdot \underline{n}, \quad \underline{\underline{\tau}} \cdot \underline{n} \cdot \underline{t} = \underline{\nabla}\sigma \cdot \underline{t} \tag{1.2d}$$

where $\underline{n}$ and $\underline{t}$ are unit vectors, normal (outward-pointing) and tangential to the interface, respectively and $\underline{\underline{\tau}} = -p\underline{\underline{I}} + 2\mu\underline{\underline{e}}$ the stress tensor with $\underline{\underline{e}}$ the rate-of-strain tensor given by $e_{ij} = (1/2)(\partial u_i/\partial x_j + \partial u_j/\partial x_i)$. p_a is the pressure of the ambient gas phase and σ is the surface tension. The thermal boundary condition on the free surface is:

$$\lambda\underline{\nabla}T \cdot \underline{n} = -\alpha_g(T - T_a) \quad \text{on}\quad y = h \tag{1.2e}$$

where α_g is the heat transfer coefficient between the liquid and air. Finally, the thermo-capillary effect is modeled by using a linear approximation for the surface tension,

$$\sigma = \sigma_a - \gamma(T - T_a) \tag{1.3}$$

with σ_a the surface tension at the reference temperature T_a, taken to be the temperature of the ambient gas phase, and $\gamma > 0$ for typical liquids (see also chapter by Thiele in this book).

The above system of Navier-Stokes and energy equations and wall and free-surface boundary conditions has a trivial solution corresponding to the plane-parallel base state

$$h = h_{\mathrm{N}}, \ p = p_a + \rho(h_{\mathrm{N}} - y)g\cos\beta, \ u = \frac{g\sin\beta}{2\nu}(2h_{\mathrm{N}}y - y^2), \ v = 0, \tag{1.4a}$$

$$\text{HF:}\ T = T_a + \beta_T[\alpha_g(h_{\mathrm{N}} - y) + \lambda], \quad \text{ST:}\ T = T_a + \hat{\beta}_T[\alpha_g(h_{\mathrm{N}} - y) + \lambda] \tag{1.4b}$$

where $\beta_T = q_0/[\lambda(\alpha_w + \alpha_g) + \alpha_w\alpha_g h_{\mathrm{N}}]$ and $\hat{\beta}_T = (T_w - T_a)/(\alpha_g h_{\mathrm{N}} + \lambda)$. The Nusselt flat film solution (1.4) corresponds to a balance of viscous and gravitational forces and so we introduce a non-dimensionalization based on the length and time scales built from the stream-wise gravity acceleration and the kinematic viscosity: $l_0 = \nu^{2/3}(g\sin\beta)^{-1/3}$ and $t_0 = \nu^{1/3}(g\sin\beta)^{-2/3}$. We then scale velocities with l_0/t_0 whilst pressure and temperature are made dimensionless using $p \to p_a + \rho(l_0/t_0)^2 p$ and $T \to T_a + T_0 T$. In the case of an imposed temperature on the wall (ST), we take $T_0 = T_w - T_a$, i.e. the temperature difference between air and wall, a natural control parameter in a real experiment, whereas when a constant heat flux is applied (HF) we take $T_0 = q_0 l_0 / \lambda$, a natural scale for the temperature.

In terms of these nondimensional variables, the equations of motion and energy equation become

$$
\begin{aligned}
u_x + v_y &= 0 & \text{(1.5a)} \\
u_{yy} + 1 &= p_x + u_t + u u_x + v u_y - u_{xx} & \text{(1.5b)} \\
p_y + \cot \beta &= v_{yy} - v_t - u v_x - v v_y + v_{xx} & \text{(1.5c)} \\
T_{yy} &= Pr(T_t + u T_x + v T_y) - T_{xx} & \text{(1.5d)}
\end{aligned}
$$

with $Pr = \nu / \kappa$ the Prandtl number. The wall boundary conditions become

$$
u = v = 0, \quad \text{HF: } T_y = -1 + Bi_w T, \quad \text{ST: } T = 1 \tag{1.6a}
$$

where $Bi_w = \alpha_w l_0 / \lambda = \alpha_w \nu^{2/3} (g \sin \beta)^{-1/3} / \lambda$ is the wall Biot number, while the interfacial boundary conditions at $y = h$ become

$$
\begin{aligned}
h_t + u h_x - v &= 0 & \text{(1.6b)} \\
p + (Ka - MaT)N^{-\frac{3}{2}} h_{xx} &= 2N^{-1}(v_y - h_x u_y + h_x^2 u_x - h_x v_x) & \text{(1.6c)} \\
u_y + Ma(T_x + h_x T_y)N^{\frac{1}{2}} &= -v_x - 2h_x(v_y - u_x) + h_x^2 u_y + h_x^2 v_x & \text{(1.6d)} \\
T_y + BiTN^{\frac{1}{2}} &= h_x T_x & \text{(1.6e)}
\end{aligned}
$$

where $N = 1 + h_x^2$. For the heat flux problem,

$$
\begin{aligned}
Ma &= \frac{\gamma q_0}{\lambda \rho g l_0 \sin \beta} = \frac{\gamma q_0}{\lambda \rho (g \sin \beta)^{2/3} \nu^{2/3}}, \quad Ka = \frac{\sigma_a}{\rho g l_0^2 \sin \beta} = \frac{\sigma_a}{\rho (g \sin \beta)^{1/3} \nu^{4/3}}, \\
Bi &= \frac{\alpha_g l_0}{\lambda} = \frac{\alpha_g \nu^{2/3}}{\lambda (g \sin \beta)^{1/3}}
\end{aligned}
$$

are the Marangoni, Kapitza and surface Biot numbers, respectively. For the specified temperature problem, the equations are the same except that now the Marangoni and surface Biot numbers are defined as

$$
\hat{Ma} = \frac{\gamma(T_w - T_a)}{\rho g l_0^2 \sin \beta} = \frac{\gamma(T_w - T_a)}{\rho \nu^{4/3}(g \sin \beta)^{1/3}}, \quad \hat{Bi} = Bi.
$$

Note that throughout this study hats will be used to denote parameters associated with the ST problem.

The system of equations (1.5) and (1.6) is governed by the inclination angle β, the dimensionless Nusselt flat film thickness $\bar{h}_N = h_N / l_0$, or equivalently the Reynolds number which appears implicitly through $\bar{h}_N$, $Re = g h_N^3 \sin \beta / 3\nu^2 = \bar{h}_N^3 / 3$, and the five dimensionless groups, Ka, Ma, Pr, Bi_w and Bi for HF and the four dimensionless groups, Ka, $\hat{Ma}$, Pr and $\hat{Bi}$ for ST. Hence, a complete investigation over the entire parameter space would be impossible. However, for fixed liquid properties and inclination angle β, the Prandtl and Kapitza numbers are fixed thus reducing the number of relevant parameters by three. Therefore, for a given liquid-gas-solid system and wall heating conditions the only free parameter is the Reynolds number which is a flow control parameter, and the heated falling film problem is a one-parameter system only. If the liquid phase now is water at $25°C$ and the plane is vertical, $Ka = 3000$ and $Pr = 7$. The HF problem then has four free parameters, Re, Ma, Bi_w and Bi while the ST problem has three free parameters, Re, $\hat{Ma}$ and $\hat{Bi}$. In the absence of experimental values for the Biot number we take $\hat{Bi} = \frac{1}{10}$ – it is realistic to expect the gas to be a poor conductor. The values for the remaining parameters, Bi_w, Bi, Ma and $\hat{Ma}$ will be discussed in the next section.

By using the above non-dimensionalization which is based on viscosity and gravity, we have only one parameter, Re, that depends on $\bar{h}_N$, with the remaining parameters, $Ka, Ma, Pr, Bi_w, Bi, \hat{Ma}$ and $\hat{Bi}$ all independent of $\bar{h}_N$ and fixed for a given gas-liquid-solid system and wall heating conditions. In experiments the film thickness is modified via the flow rate and hence it is useful to have only one parameter that depends on $\bar{h}_N$. We note that the flat film solution $\bar{h}_N$ is also the boundary condition $h \to \bar{h}_N$ far from a solitary hump. However, for numerical purposes, the formulation of a model in which the film thickness has been scaled out of the boundary conditions and the flat film solution is fixed, thus allowing useful comparisons to be made, would be welcome. Hence, for convenience we employ a scaling based on the Nusselt solution through the transformation

$$x \to \bar{h}_N x, \quad (y, h) \to \bar{h}_N(y, h), \quad (u, v) \to \bar{h}_N^2(u, v), \quad t \to t/\bar{h}_N, \quad p \to \bar{h}_N p \quad (1.7a)$$

$$(HF) : T \to \bar{h}_N T, \quad (ST) : T \to T. \quad (1.7b)$$

Equations (1.5) are thus transformed into

$$u_x + v_y = 0 \tag{1.8a}$$

$$u_{yy} + 1 = p_x + 3Re(u_t + uu_x + vu_y) - u_{xx} \tag{1.8b}$$

$$p_y + \cot\beta = v_{yy} - 3Re(v_t + uv_x + vv_y) + v_{xx} \tag{1.8c}$$

$$T_{yy} = 3Pe(T_t + uT_x + vT_y) - T_{xx}, \tag{1.8d}$$

where $Pe = PrRe$, is the Péclet number. Note that the transformation (1.7) does not introduce numerical factors in the equations except for a factor of three appearing along with the Reynolds and Péclet numbers, which is related to the definition of the Reynolds number based on the flow rate. The wall boundary conditions now become

$$u = v = 0, \quad (HF): T_y = -1 + B_w T, \quad (ST): T = 1 \tag{1.9a}$$

and the free-surface boundary conditions are written as

$$h_t + uh_x - v = 0 \tag{1.9b}$$

$$p + (We - MT)N^{-\frac{3}{2}}h_{xx} = 2N^{-1}(v_y - h_x u_y + h_x^2 u_x - h_x v_x) \tag{1.9c}$$

$$u_y + M(T_x + h_x T_y)N^{\frac{1}{2}} = -v_x - 2h_x(v_y - u_x) + h_x^2 u_y + h_x^2 v_x \tag{1.9d}$$

$$T_y + BTN^{\frac{1}{2}} = h_x T_x. \tag{1.9e}$$

This scaling has modified the definitions of the Marangoni and Biot numbers via the introduction of modified parameters, M, $\hat{M}$, B, $\hat{B}$ and B_w that are based on the Nusselt flat film solution,

$$We \; = \; \frac{\sigma_a}{\rho g h_N^2 \sin \beta} = \frac{Ka}{\bar{h}_N^2}\,, \qquad M = \frac{\gamma q_0}{\lambda \rho g h_N \sin \beta} = \frac{Ma}{\bar{h}_N}\,, \qquad (1.10a)$$

$$B_w \; = \; \alpha_w h_N/\lambda = Bi_w \bar{h}_N\,, \qquad B = \alpha_g h_N/\lambda = Bi\bar{h}_N\,, \qquad (1.10b)$$

$$\hat{M} \; = \; \frac{\gamma(T_w - T_a)}{\rho g h_N^2 \sin \beta} = \frac{\hat{M}a}{\bar{h}_N^2}\,, \qquad \hat{B} = \alpha_g h_N/\lambda = \hat{B}i\bar{h}_N\,, \qquad (1.10c)$$

where We is the Weber number. Equations (1.8) and (1.9) are the basic equations for the analysis to follow.

1.3 Long-Wave Theory for Large Péclet Numbers

The set of equations in (1.8) and (1.9) is highly nonlinear and complex due to the presence of the free surface, the coupling of the temperature and velocity fields due to the Marangoni effect and the transport of heat by the flow. However, as surface tension is generally high, the observed waves should be typically long compared to the film thickness, thus justifying a long-wave assumption for an interface slowly varying in time and space. By introducing a formal parameter ϵ corresponding to a typical slope of the film, we can then perform a gradient expansion $\partial_{x,t} \sim \epsilon \ll 1$ as done initially by Benney (1966) in the case of isothermal flow (see also chapter by Thiele in this book).

A classical Benney expansion would typically assume all dimensionless parameters to be of order unity, with the exception of the Weber number which is taken to be much larger, typically $We = O(\epsilon^{-2})$. The usual trick is then to bring at order ϵ the dominant surface tension effects. These stabilizing terms prevent the waves from breaking thus satisfying the long-wave approximation. However, as a consequence of an order unity Péclet number, the convective heat transport effects are retained if one expands the velocity field up to $O(\epsilon^2)$ and the temperature field up to $O(\epsilon)$ but these higher order corrections are rather lengthy.

In practice the Péclet number can be much larger than the Reynolds number due to the ratio of the momentum and thermal diffusivities being much larger than unity – recall e.g. that for water the Prandtl number is 7. We then expect that convection at large Péclet numbers can lead to a downstream convective distortion of the free-surface temperature distribution obtained by assuming an order unity Péclet number. As a result the transport of heat by the flow becomes important which then significantly modifies the interfacial temperature and consequently the Marangoni effect. Hence we assume $Pe \sim O(\epsilon^{-n})$ for $0 < n < 1$ such that the convective heat transport effects are included at a lower relevant order and transport of heat by the flow becomes significant in the heat balance.

We then carry out an expansion up to $O(\epsilon^{2-n})$ and we neglect terms of $O(\epsilon^2)$ and higher, thus removing the complexity of the free-boundary problem in (1.8) and (1.9). This level of truncation allows the derivation of a relatively simple evolution equation for the local film thickness. The pressure and temperature are both expanded up to $O(\epsilon^{1-n})$

and hence terms of $O(\epsilon)$ and higher are omitted in these expansions. At this level of truncation, the pressure is given by

$$p \;=\; (h-y)\cot\beta - Weh_{xx} \tag{1.11}$$

and the temperature field is given by

$$
\begin{aligned}
\text{HF: } T \;=\; & [1+B(h-y)]F - PeBB_w F^2 h_x y^4 \left(\frac{3By}{40} - \frac{1}{8} - \frac{3B}{8}h\right) \\
& - PeB^2 F^2 hh_x y^2 \left[(B_w h - 1)\frac{y}{2} + \frac{3h}{2}\right] \\
& + PeB(1+B_w y)F^3 h^3 h_x \left[\frac{B^2}{5}h(5+B_w h) + \frac{B_w}{4}(Bh-2) + \frac{3}{2}B\right] \tag{1.12a}
\end{aligned}
$$

$$
\begin{aligned}
\text{ST: } T \;=\; & (1+\hat{B}(h-y))\hat{F} - Pe\hat{B}\hat{F}^2 h_x y^3 \left[\frac{3\hat{B}}{40}y^2 - \left(\frac{1}{8} + \frac{3\hat{B}}{8}h\right)y + \frac{\hat{B}}{2}h^2\right] \\
& + Pe\hat{B}\hat{F}^3 h^3 h_x y \left(\frac{\hat{B}^2}{5}h^2 + \frac{\hat{B}}{4}h - \frac{1}{2}\right), \tag{1.12b}
\end{aligned}
$$

where $F = (B+B_w+BB_w h)^{-1}$ and $\hat{F} = (1+\hat{B}h)^{-1}$. Notice that taking the leading order terms from the limit of infinite B_w in equation (1.12a) yields equation (1.12b) except for the $1/B_w$ factor on the right hand side. The velocity components now are obtained from $u = \psi_y$ and $v = -\psi_x$ where the streamfunction ψ is given by

$$
\begin{aligned}
\text{HF: } \psi \;=\; & y^2 \left(\frac{h}{2} - \frac{y}{6}\right)(1 + Weh_{xxx} - h_x \cot\beta) + hh_x y^2 \frac{Re}{40}(20h^3 - 5hy^2 + y^3) \\
& + \frac{M}{2}BB_w F^2 h_x y^2 - \frac{Pe}{80}MB\left(Gh^3 h_x\right)_x y^2 \tag{1.13}
\end{aligned}
$$

where $G = 7(4+B_w h)F^2 + (32B - 22B_w^2 h - 48B_w)F^3$. For ST the streamfunction is given by (1.13) where G, $MBB_w F^2$ and MB are replaced by $\hat{G} = (7\hat{B}h - 15)h\hat{F}^3$, $\hat{M}\hat{B}\hat{F}^2$ and $\hat{M}\hat{B}$, respectively. The free-surface evolution equation can then be easily obtained from the kinematic boundary condition in (1.9b):

$$
\begin{aligned}
\text{HF: } h_t \;+\; & h^2 h_x + \left(\frac{2}{5}Reh^6 h_x - \frac{1}{3}h^3 h_x \cot\beta + \frac{1}{2}MBB_w F^2 h^2 h_x + \frac{1}{3}Weh^3 h_{xxx}\right)_x \\
& - \frac{Pe}{80}MB\left[h^2 \left(Gh^3 h_x\right)_x\right]_x = 0. \tag{1.14}
\end{aligned}
$$

The equivalent evolution equation for the ST problem can then be obtained by simply replacing G, $MBB_w F^2$ and MB by $\hat{G}$, $\hat{M}\hat{B}\hat{F}^2$ and $\hat{M}\hat{B}$, respectively. The second term in (1.14) is the convective term due to mean flow, the third term arises from inertia, the fourth term is due to the inclination of the plane, the fifth term arises from the Marangoni effect and is responsible for the Marangoni instability (for $M > 0$), the sixth term is the streamwise curvature gradient associated with surface tension and the seventh term originates from the heat transport convective terms. The first two terms are of $O(1)$, the third, fourth, fifth and sixth terms are of $O(\epsilon)$ and the seventh term is of $O(\epsilon^{2-n})$.

A linear stability analysis of the trivial solution $h = 1$ of (1.14) and the equivalent evolution equation for ST yields the critical conditions

$$\text{HF: } Re_c \;=\; \frac{5}{6}\cot\beta - \frac{5}{4}MBB_w\overline{F}^2 \tag{1.15a}$$

$$\text{ST: } Re_c \;=\; \frac{5}{6}\cot\beta - \frac{5}{4}\hat{M}B\overline{\hat{F}}^2 \tag{1.15b}$$

where Re_c is the critical Reynolds number above which the flow looses stability and $\overline{F} = F|_{h=1}$, $\overline{\hat{F}} = \hat{F}|_{h=1}$. These conditions indicate that for $M, \hat{M} > 0$, the Marangoni effect is destabilizing as Re_c decreases with increasing $M, \hat{M}$. On the other hand for $M, \hat{M} < 0$, the Marangoni effect is stabilizing as Re_c increases with increasing $|M|, |\hat{M}|$. For $M = \hat{M} = 0$, the above expressions reduce to the well-known critical condition for a free-falling film, $Re_c = (5/6)\cot\beta$ (Benjamin, 1957; Yih, 1963). Note that (1.15) is identical to the criticality conditions obtained from the Orr-Sommerfeld eigenvalue problem of the full Navier-Stokes and energy equation (Trevelyan et al., 2006). This is not surprising as the long-wave expansion is a regular perturbation expansion of the full Navier-Stokes and energy equations and should be exact close to criticality.

We also note that for the HF problem with $B_w = 0$ the fifth term in (1.14) responsible for the Marangoni instability vanishes and in this case the Marangoni effect only contributes to the dispersion of the waves through the last term in (1.14), but it does not influence the instability onset. This means that for a specified heat flux boundary condition, or equivalently a plate that is perfectly insulated from the gas phase below, the long-wave thermocapillary instability is suppressed. In this case, the interfacial temperature distribution is $T|_{y=h} = B^{-1} + (3/2)PeB^{-1}h^3h_x$ and has two contributions: B^{-1} due to heat conduction across the film and $(3/2)PeB^{-1}h^3h_x$ due to convective heat transport. The first term is independent of h and as a result thermocapillarity does not affect the instability as variations of h do not induce perturbations on the interfacial temperature distribution through heat conduction – see also Scheid et al. (2002) for a discussion of the specified heat flux boundary condition.

On the other hand, for the ST problem heat conduction contributes the term $(1 + \hat{B}h)^{-1}$ in $T_{y=h}$ and the Marangoni forces in this case always influence the onset of the instability. However, if $\hat{B} = 0$ for the ST problem, i.e. the interface is a poor heat conductor perfectly insulated from the surrounding gas, the Marangoni effect does not influence the system. In this case $T = 1$ from (1.12b) and the temperature is everywhere uniform and equal to the wall temperature so that there is no instability due to the thermal effects or influence on the dispersion of the waves and the momentum and heat transport problems are decoupled.

For convenience let us now rescale the evolution equations using the scalings introduced by Shkadov (1977). This author introduced a length scale in the streamwise direction corresponding to the balance of the streamwise pressure gradient $\sigma_a h_{xxx}$ due to surface tension and the streamwise gravity acceleration $\rho g \sin\beta$. This length scale, say l_S, corresponds effectively to the characteristic size of the steep front of the waves.

Note that l_S should be much larger than the film thickness $h_{\rm N}$ in order to sustain the long-wave assumption. Simple algebra then shows that $l_S/h_{\rm N} = We^{1/3}$ which is a large ratio as long as the Weber number is sufficiently large. The rescaled space and time coordinates are then defined as $x = We^{1/3}X$ and $t = We^{1/3}\Theta$ to yield

$$\text{HF: } h_\Theta + h^2 h_X + \left(\mathcal{A}(h)h_X + \mathcal{B}(h)h_X^2 + \mathcal{C}(h)h_{XX} + \frac{1}{3}h^3 h_{XXX} \right)_X = 0 \qquad (1.16)$$

where $\mathcal{A}(h) = \frac{2}{15}\delta h^6 - \frac{1}{3}h^3\zeta + \frac{1}{2}\mathcal{M}BB_w F^2 h^2$, $\mathcal{B}(h) = -\frac{1}{240}Pr\delta\mathcal{M}Bh^4\left(3G + h\frac{\partial G}{\partial h}\right)$ and $\mathcal{C}(h) = -\frac{1}{240}Pr\delta\mathcal{M}BGh^5$. For (ST) we obtain a similar equation where the functions $\mathcal{A}(h)$, $\mathcal{B}(h)$ and $\mathcal{C}(h)$ are now replaced by $\hat{\mathcal{A}}(h) = \frac{2}{15}\delta h^6 - \frac{1}{3}h^3\zeta + \frac{1}{2}\hat{\mathcal{M}}\hat{B}F^2 h^2$, $\hat{\mathcal{B}}(h) = -\frac{1}{240}Pr\delta\hat{\mathcal{M}}\hat{B}h^4\left(3\hat{G} + h\frac{\partial \hat{G}}{\partial h}\right)$ and $\hat{\mathcal{C}}(h) = -\frac{1}{240}Pr\delta\hat{\mathcal{M}}\hat{B}\hat{G}h^5$. $\delta = 3Re/We^{1/3}$, $\mathcal{M} = M/We^{1/3}$ and $\hat{\mathcal{M}} = \hat{M}/We^{1/3}$ are reduced Reynolds and Marangoni numbers, respectively, and $\zeta = \cot\beta/We^{1/3}$ is a reduced slope. Notice that, compared to the definition of the reduced Reynolds number defined by Shkadov, δ is 45 times larger, this ratio originating from a slightly different choice of the scalings with the aim being to avoid the introduction of numerical factors that may complicate the conversion between the different scalings used in this study.

In what follows, the evolution equation for the film thickness h in (1.16) will be referred to as LWE-HF and the corresponding evolution equation for ST as LWE-ST. These long-wave expansions have been obtained to check the behavior of the models developed in the next section close to criticality. It is also exactly because of the presence of the convective heat transport terms in these models, that we have developed a long-wave theory to include these terms. Finally, the LWE expansions developed here motivates our choice of values for the parameters Bi_w, Bi, Ma and $\hat{M}a$ used later on in our computations. More specifically we choose two sets of parameter values, $Bi_w = Bi = \frac{1}{5}$ with $Ma = 2Bi\hat{M}a$ and $Bi_w = 5Bi = \frac{3}{5}$ with $Ma = 6Bi\hat{M}a$. Ma is chosen so that in the absence of the heat transport convective heat effects (equivalently for small Pe) $\mathcal{C} = \hat{\mathcal{C}} = 0$ and the two equations LWE-ST and LWE-HF are identical. For LWE-ST, $\hat{\mathcal{C}} > 0$ for $\hat{B}h < 15/7$ which is always satisfied while for LWE-HF and for sufficiently small film thicknesses (close to criticality), $\mathcal{C} < 0$ for $Bi_w < 3Bi$ (this is the case with the first set of parameters for Bi, Bi_w) and $\mathcal{C} > 0$ for $Bi_w > 3Bi$ (this is the case with the second set of parameters for Bi, Bi_w) (more details are given in Trevelyan et al. (2006)). The functional form of the LWE approximations developed here is similar to the LWE approximation for the problem of a reactive falling film considered recently by Trevelyan and Kalliadasis (2004a,b). These authors demonstrated that sufficiently close to criticality, the sign of $\mathcal{C}$ (which is also the sign of dispersion) affects the type of solitary waves with $\mathcal{C} > 0$ leading to positive-hump waves while $\mathcal{C} < 0$ leading to negative-hump waves. Hence, we anticipate that the second set of parameter values will give a better qualitative agreement between HF and ST than the first set (at least close to criticality) due to the change in sign for $\mathcal{C}$.

1.4 Weighted Residuals Approach

The LWE developed in the previous section leads to a single evolution equation for the film thickness h that contains high-order nonlinearities. In the case of isothermal film flows, it has been shown that these nonlinearities are responsible for the presence of finite-time blow-up behavior of the non-stationary problem (Pumir et al., 1983; Scheid et al., 2005b). Due to this unphysical behavior, the long-wave approach is actually limited to a narrow range of parameters around the onset of the instability. One reason for this failure is certainly the assumed slaving of the velocity field to the movement of the interface, thus leading to a single evolution equation for the film thickness h. A possible way out would therefore be to introduce more degrees of freedom and turn to models in terms of systems of coupled evolution equations for several fields.

Such coupled systems of equations can be easily obtained by averaging the basic equations across the fluid layer. This process enables us to turn from the description of the motion of a fluid particle to the motion of a column of fluid from $y = 0$ to $y = h$. Obviously, such an approximation is valid only if a strong coherence between different layers of fluid from $y = 0$ to $y = h$ exists. Such a coherence can be sustained only by viscosity and therefore must result from the long-wave assumption. Hence, the starting point of our analysis is still to assume long waves in the streamwise direction. For simplicity, we shall also neglect the second order diffusive terms u_{xx} and T_{xx} of the Navier-Stokes and energy equations.

To leading order, the y-component of the equation of motion (1.8c) and normal stress balance (1.9c) are $p_y = -\cot\beta$ and $p = -We h_{xx}$ on $y = h(x,t)$. Hence, we obtain the same leading-order pressure distribution as in equation (1.11). Substituting now the expression for the pressure into the x-component of the momentum equation (1.8b) and neglecting terms of $O(\epsilon^2)$ and higher yields

$$u_{yy} + 1 = h_x \cot\beta - We h_{xxx} + 3Re(u_t + uu_x + vu_y). \qquad (1.17a)$$

The y-component of the velocity can be eliminated using the continuity equation (1.8a) along with the no-slip condition (1.9a), to obtain $v = -\int_0^y u_x dy'$. The u-velocity must satisfy the no-slip boundary condition and the leading-order tangential stress balance on the interface from (1.9d):

$$u = 0 \quad \text{on } y = 0, \quad u_y = -M\theta_x \quad \text{on} \quad y = h, \qquad (1.17b)$$

where terms of $O(\epsilon^2)$ and higher have been neglected and $\theta(x,t)$ is the interfacial temperature, i.e. $\theta \equiv T|_{y=h}$ and $\theta_x \equiv (T_x + h_x T_y)|_{y=h}$. The above system is coupled with the energy equation and thermal boundary conditions, however, we can examine the flow field by assuming that the function θ is known. The system is then closed via the kinematic condition:

$$h_t + uh_x = v \quad \text{on} \quad y = h.$$

By integrating the continuity equation $u_x + v_y = 0$ across the film, this last condition can be written as

$$h_t + q_x = 0 \qquad (1.17c)$$

where $q = \int_0^h u\, dy$ is the flow rate. In the absence of the Marangoni term $M\theta_x$ appearing in the stress balance at the free surface, equations (1.17) are the so-called 'boundary-layer equations'.

Let us now expand the unknown velocity field on a set of $n+1$ test functions f_i as $u(x,y,t) = \sum_{i=0}^{n} A_i(x,t) f_i(\eta)$ where $\eta = y/h(x,t)$ is a reduced normal coordinate. Introduction of this ansatz into (1.17a) yields the residual

$$R_u = 3Re(u_t + uu_x + vu_y) - u_{yy} - 1 + h_x \cot\beta - We h_{xxx}. \tag{1.18}$$

The weighted residuals approach then requires that we cancel appropriately weighted integrals of the residual, namely,

$$\langle w_i, R_u \rangle = 0, \tag{1.19}$$

where w_i are weight functions and the inner product is defined as $\langle f, g \rangle = \int_0^h fg\,dy$ for any two functions f and g with appropriate boundary conditions. We then obtain coupled equations for the $n+1$ amplitudes A_i. Notice that replacing the condition $R_u = 0$ from (1.18) with the averaged equation (1.19) reduces the number of spatial variables from two to one.

Due to the complexity of the system in (1.19), the number of test functions is usually reduced to only one as was done by Shkadov (1967, 1968) for isothermal flows where a uniform weight equal to unity was applied whilst Usha and Uma (2004) in their study of isothermal flows chose the weight function to be exactly the same with the test function; this is effectively a Galerkin projection with just a single test function (the Galerkin projection for isothermal flows was first suggested by Ruyer-Quil and Manneville (2000, 2002)). Note that when the expansion of the velocity field is restricted to only one test function, the velocity field has a self-similar form such that for two different locations x_1 and x_2 on the plane we have

$$\frac{u(x_1, \eta h(x_1))}{u(x_1, h(x_1))} = \frac{u(x_2, \eta h(x_2))}{u(x_2, h(x_2))}.$$

For isothermal flows, the following self-similar parabolic profile is commonly assumed,

$$u = 3\frac{q}{h}\left(\eta - \frac{1}{2}\eta^2\right), \tag{1.20}$$

which corresponds to the Nusselt flat film flow. The basic assumption here is that a parabolic velocity profile which satisfies the x-component of the equation of motion for zero Reynolds number persists even for moderate Reynolds numbers when the free surface is no longer flat. The resulting averaged equation is the Shkadov IBL approximation for isothermal flows. In the presence of the Marangoni effect, (1.20) does not satisfy the tangential stress balance. In this case, Kalliadasis et al. (2003a,b) proposed the test function,

$$u = 3\frac{q}{h}\left(\eta - \frac{1}{2}\eta^2\right) + M\theta_x h\left(\frac{1}{2}\eta - \frac{3}{4}\eta^2\right), \tag{1.21}$$

which is the simplest possible velocity profile satisfying all boundary conditions.

The averaging approaches developed by Shkadov (1967, 1968) and Usha and Uma (2004) for isothermal flows are effectively special cases of the n-th moment of R_u, namely,

$$
\langle u^n, R_u \rangle = \frac{3Re}{n+1} \left(\int_0^h u^{n+1} dy \right)_t + \frac{6Re}{n+1} \left(\int_0^h u^{n+2} dy \right)_x - \int_0^h u^n u_{yy} dy
$$
$$
+ \quad (1 - h_x \cot \beta + We h_{xxx}) \int_0^h u^n dy. \tag{1.22}
$$

For non-isothermal flows, using the test function (1.21) and truncating R_u at order ϵ yields

$$
\tilde{R}_u = 3Re(u_t^{(0)} + u^{(0)} u_x^{(0)} + v^{(0)} u_y^{(0)}) - u_{yy} - 1 + h_x \cot \beta - We h_{xxx} \tag{1.23}
$$

where $u^{(0)}$ denotes the leading-order term from u (which is also the same with (1.20)) and $v^{(0)} = - \int_0^y u_x^{(0)} dy'$. Indeed, the Marangoni terms in (1.21) are of $O(\epsilon M)$ so that they only contribute to the viscous diffusion term $\partial^2/\partial y^2$ and are neglected in the inertial/convective terms which are of $O(\epsilon Re)$.

Setting $n = 0$ in (1.22) and retaining the dominant terms yields the inner product $\langle 1, \tilde{R}_u \rangle$ which is the Shkadov IBL approach for non-isothermal flows. The same weight function was also adopted by Kalliadasis et al. (2003a,b) in the presence of thermal effects. The resulting averaged momentum equation is,

$$
3Re \left[q_t + \frac{6}{5} \left(\frac{q^2}{h} \right)_x \right] + \frac{3q}{h^2} = h + We h h_{xxx} - h h_x \cot \beta - \frac{3}{2} M \theta_x,
$$

an evolution equation for q which also involves the interfacial temperature θ (which is unknown at this stage). On the other hand, setting $n = 1$ in (1.22) and retaining the dominant terms yields the inner product $\langle u^{(0)}, \tilde{R}_u \rangle$ or equivalently,

$$
\frac{18}{5} Re \left(q_t + \frac{17}{7} \frac{q}{h} q_x - \frac{9}{7} \frac{q^2}{h^2} h_x \right) + \frac{3q}{h^2} = h + We h h_{xxx} - h h_x \cot \beta - \frac{3}{2} M \theta_x, \tag{1.24}
$$

which is the momentum equation used in the remainder of this study. Note that equations (1.17c) and (1.24) correspond to equations (5.17a) and (5.17b) in Scheid (2004).

For the isothermal falling film problem, Ruyer-Quil and Manneville (2000, 2002) developed high-order IBL models using refined polynomial expansions for the velocity field [corresponding to corrections of the Shkadov parabolic self-similar profile; evidently deviations to (1.20) are necessarily produced by modulations of the free surface and are at least of order ϵ] and high-order weighted residuals techniques including the Galerkin projection (in which the weight functions are equal to the test functions). Hence, equation (1.24) is the Ruyer-Quil and Manneville single-test function Galerkin approach but for non-isothermal flows. For the isothermal case, Ruyer-Quil and Manneville showed that the single-test function Galerkin projection fully corrects the critical Reynolds number obtained from the Shkadov IBL approximation. We shall demonstrate that this is also the case in the presence of Marangoni effects, in fact it is only necessary to take the weight function as the leading order test function for the velocity, namely $w_u \equiv \eta - \frac{1}{2}\eta^2$.

Simple weighted residuals approach for the energy equation. By solving equations (1.17c) and (1.24) for h and q, the velocity field can be obtained from (1.21), provided of course that θ is known. In this section we outline a simple approach to obtain a single equation for θ. The wall boundary condition for HF is given in (1.9a), i.e. $T_y = B_w T - 1$, while the leading-order surface boundary condition from (1.9e) is

$$T_y = -BT \quad \text{on} \quad y = h \tag{1.25}$$

where terms of $O(\epsilon^2)$ and higher have been neglected. Like with the momentum equation, the first step for the energy equation would be the introduction of a self-similar profile. The flat film solution has a linear temperature distribution. We can then choose a linear profile which satisfies the wall boundary condition in (1.9a) along with $T|_{\eta=1} = \theta$ to obtain

$$T = \theta + \frac{1 - B_w\theta}{1 + B_w h}h(1 - \eta) \tag{1.26a}$$

for HF and

$$T = 1 + (\theta - 1)\eta. \tag{1.26b}$$

for ST. Hence, the assumption here is that the linear temperature profile obtained for a flat film persists even when the interface is no longer flat. Note that θ_x occurs explicitly in the momentum equation (1.24) and so it is convenient to explicitly include θ in the temperature fields. It is also important to note that equation (1.26a) yields $T = F[1 + Bh(1 - \eta)]$ when $\theta = F$. Similarly, equation (1.26b) yields $T = \hat{F}[1 + \hat{B}h(1 - \eta)]$ when $\theta = \hat{F}$. Thus, these expressions are consistent with the flat film temperature distributions.

By analogy now with our analysis for the momentum equation, the introduction of the above test functions for the temperature fields yields a residual for the energy equation at $O(\epsilon)$

$$R_T = 3Pe(T_t + u^{(0)}T_x + v^{(0)}T_y) - T_{yy} \tag{1.27}$$

where the terms of $O(\epsilon M)$ of u and v contribute only to the thermal diffusion term $\partial^2/\partial y^2$ and are neglected in the inertial/convective terms which are of $O(\epsilon Pe)$. The energy residuals can then be minimized from

$$\langle w_T, R_T \rangle = 0 \tag{1.28}$$

where w_T is an appropriately chosen weight function. We note that although the temperature distributions in (1.26a) and (1.26b) satisfy their respective wall boundary conditions in (1.9a), they do *not* satisfy the interfacial condition (1.25), unlike the velocity profile in (1.21) which satisfies all boundary conditions. It is in fact impossible for a linear profile to satisfy (1.25) and (1.28), however, as was pointed out by Kalliadasis et al. (2003a) by choosing the weight function appropriately, the boundary terms resulting from integrations by parts involve either T_η on $\eta = 1$ or T on $\eta = 0$ and thus the interfacial boundary condition can be included in the boundary terms resulting from the integrations by parts. Hence, although the test function does not satisfy all boundary conditions, the averaged energy equation does and the flat film solution can still be retained in our averaging formulation.

For HF we take $w_T \equiv 1$ so that $\langle w_T, T_{yy} \rangle$ becomes

$$\int_0^h T_{yy}\,dy = [T_y]_0^h = -B\theta - T_y|_{\eta=0} = -B\theta + \frac{1 - B_w\theta}{1 + B_w h} = \frac{1 - \theta F^{-1}}{1 + B_w h}$$

where the surface boundary condition has been substituted for $T_y|_{y=h}$ and the profile in (1.26a) was used in $T_y|_{y=0}$. Using (1.26a) for the convective terms along with the above expression for $\langle w_T, T_{yy} \rangle$ we can evaluate (1.28) to obtain:

$$
\begin{aligned}
0 &= \frac{2 + B_w h}{2}\theta_t + \frac{8 + 5B_w h}{8h}q\theta_x - \frac{(1 - B_w\theta)}{8(1 + B_w h)}\left[(5 + B_w h)q_x - 3\frac{qh_x}{h}\right] \\
&\quad + \frac{\theta F^{-1} - 1}{3Peh}.
\end{aligned}
$$

(1.29a)

Equations (1.24) and (1.29a) along with the kinematic boundary condition in (1.17c) will be referred to as the SHF model – a simple heat flux model.

For ST we take $w_T \equiv y$ so that $\langle w_T, T_{yy} \rangle$ becomes

$$\int_0^h y T_{yy}\,dy = [y T_y]_0^h - \int_0^h T_y\,dy = h T_y|_{y=h} - T|_{y=h} + T|_{y=0} = -\hat{B}h\theta - \theta + 1$$

where both the interfacial boundary condition and wall boundary condition have been used within the integrations by parts. If this had not been done, $\langle w_T, T_{yy} \rangle$ would have yielded zero. Using (1.26b) for the convective terms along with the above term for $\langle w_T, T_{yy} \rangle$ we can evaluate (1.28) to obtain:

$$0 = \theta_t + \frac{27q}{20h}\theta_x + \frac{7q_x(\theta - 1)}{40h} + \frac{1}{Peh^2}(\theta\hat{\Gamma}^{-1} - 1)$$

(1.29b)

Equations (1.24) and (1.29b) along with the kinematic boundary condition in (1.17c) will be referred to as the SST model – a simple specified temperature model. Equations (1.29a) and (1.29b) correspond to equations (8.8) and (5.17c) in Scheid (2004).

For consistency the SHF and SST models are both rescaled in the same way with the LWE-HF and LWE-ST, i.e. $x = We^{1/3}X$ and $t = We^{1/3}\Theta$ and all computations performed in this study use these scaled coordinates, namely, X and Θ.

Galerkin approach for the energy equation. The simple weighted residuals models SHF and SST are useful prototypes for the study of the dynamics of a heated film. Moreover, a linear stability of these models show that they do predict the critical Reynolds number given in (1.15). However, we also wish to obtain close to criticality the corresponding LWE models.

A more refined treatment of the temperature field will enable a weighted residuals approach to yield LWE via an appropriate gradient expansion. We consider a general polynomial expansion for the temperature field in powers of η and whose amplitudes are only functions of x and t,

$$T = \sum_{i=-1}^{m} A^{(i-1)}(x,t)\eta^i; \quad T = \sum_{i=-2}^{m} A^{(i-2)}(x,t)\eta^i,$$

for the HF and ST problems, respectively. Note that unlike the studies by Scheid (2004) and Ruyer-Quil et al. (2005) where the amplitudes in the expansion for the temperature field are assigned certain orders with respect to ϵ, in our projection for the temperature field the order of the amplitudes is not specified. Note also that the test functions utilized in Scheid (2004) and Ruyer-Quil et al. (2005) did not satisfy all boundary conditions; instead these authors chose the weight functions appropriately, so that the boundary condition can be included in the boundary terms resulting from the integrations by parts, as was done in the previous subsection.

Here we require that the temperature field satisfies all of its boundary conditions along with $T|_{\eta=1} = \theta$, a total of three conditions that need to be satisfied. For the HF problem then we eliminate a total of three amplitudes, $A^{(-1)}$, $A^{(0)}$ and $A^{(1)}$. For the ST problem we also utilize the condition $T_{yy} = 0$ on the wall which originates from a Taylor series expansion of the energy equation (1.8d) at $y = 0$ (this is also consistent with LWE-ST; T in (1.12b) has no quadratic term in y) and hence we eliminate four amplitudes, $A^{(-2)}$, $A^{(-1)}$, $A^{(0)}(\equiv 0)$ and $A^{(1)}$. In weighted residuals terminology, the elimination of these amplitudes for the HF and ST problems is effectively equivalent to a 'tau' method – see page 172 in Gottlieb and Orszag (1977).

We then project the temperature field onto the new sets of test functions $\phi, \hat{\phi}_i$,

$$\text{HF: } T = \phi_0(\eta) + \theta\phi_1(\eta) + \sum_{i=2}^{m} A^{(i)}(x,t)\phi_i(\eta) \tag{1.30a}$$

$$\text{ST: } T = \hat{\phi}_0(\eta) + \theta\hat{\phi}_1(\eta) + \sum_{i=2}^{m} A^{(i)}(x,t)\hat{\phi}_i(\eta) \tag{1.30b}$$

where we now have m amplitude functions for both cases, θ, $A^{(2)}...A^{(m)}$ and the test functions are defined by

$$\phi_0 = \frac{h(1-\eta)^2}{2 + B_w h} \,, \quad \phi_1 = 1 + (1-\eta)\eta Bh + \phi_0(B - B_w) \,,$$

$$\phi_i = \eta^{i+1} - (i+1)\frac{\eta^2}{2} + \frac{(i-1)(2 + (2-\eta)\eta B_w h)}{2(2 + B_w h)} \quad \text{for} \quad 2 \le i \le m$$

and

$$\hat{\phi}_0 = 1 - \frac{3\eta}{2} + \frac{\eta^3}{2} \,, \quad \hat{\phi}_1 = 1 - \hat{\phi}_0 + \hat{B}h(1-\eta^2)\frac{\eta}{2} \,,$$

$$\hat{\phi}_i = (i-1)\frac{\eta}{2} - (i+1)\frac{\eta^3}{2} + \eta^{i+2} \quad \text{for} \quad 2 \le i \le m.$$

Note that the functions ϕ_i and $\hat{\phi}_i$ satisfy the same conditions on the free surface, namely, $\phi_0 = \hat{\phi}_0 = \phi_1 - 1 = \hat{\phi}_1 - 1 = \phi_i = \hat{\phi}_i = 0$ and $\phi_{0_\eta} = \hat{\phi}_{0_\eta} = \phi_{1_\eta} + Bh = \hat{\phi}_{1_\eta} + \hat{B}h = \phi_{i_\eta} = \hat{\phi}_{i_\eta} = 0$. On the wall the functions ϕ_i satisfy $\phi_{0_\eta} + h - B_w h \phi_0 = \phi_{1_\eta} - B_w h \phi_1 = \phi_{i_\eta} - B_w h \phi_i = 0$, while the functions $\hat{\phi}_i$ satisfy $\hat{\phi}_0 - 1 = \hat{\phi}_1 = \hat{\phi}_i = 0$, thus 'homogenizing' the inhomogeneous wall boundary conditions. Note also that all the ϕ_i's and $\hat{\phi}_i$'s are non-negative inside the open interval $(0, 1)$.

These new sets of test functions ϕ_i and $\hat{\phi}_i$ allow the temperature fields in (1.30a) and (1.30b) to satisfy all their boundary conditions for each problem (unlike the linear profiles in the previous subsection which cannot satisfy the interfacial boundary condition) as well as to explicitly include θ.

We now let w_j denote the weight functions for the energy equation. The dominant terms from the residuals $\langle w_j, R_T \rangle = 0$ are then denoted by r_j. In the Galerkin weighted residuals approach, we choose our weight functions to be identical to our test functions, i.e. $w_j \equiv \phi_j$. Thus we obtain the set of m equations for HF:

$$r_j = 3Pe\left(\alpha_{1j}\theta_t + \beta_{1j}\theta_x + \gamma_{1j}\theta + \delta_j + \sum_{i=2}^{m}\alpha_{ij}A_t^{(i)} + \beta_{ij}A_x^{(i)} + \gamma_{ij}A^{(i)}\right)$$
$$- \Delta_j - \Gamma_{1j}\theta - \sum_{i=2}^{m}\Gamma_{ij}A^{(i)} \quad \text{for} \quad 1 \le j \le m \tag{1.31}$$

where $\alpha_{ij} = \langle \phi_j, \phi_i \rangle$, $\beta_{ij} = \langle \phi_j, u^{(0)}\phi_i \rangle$, $\gamma_{ij} = \langle \phi_j, \phi_{it} + u^{(0)}\phi_{ix} + v^{(0)}\phi_{iy} \rangle$, $\delta_j = \langle \phi_j, \phi_{0t} + u^{(0)}\phi_{0x} + v^{(0)}\phi_{0y} \rangle$, $\Delta_j = \langle \phi_j, \phi_{0yy} \rangle$ and $\Gamma_{ij} = \langle \phi_j, \phi_{iyy} \rangle$. The corresponding equations for ST are obtained from (1.31) by simply replacing ϕ_j with $\hat{\phi}_j$.

At this point it is convenient to use matrix notation, and so by introducing $\underline{A} = [\theta, A^{(2)}...A^{(m)}]^t$ we have

$$3Pe\left(\underline{\underline{M_\alpha}}\,\underline{A}_t + \underline{\underline{M_\beta}}\,\underline{A}_x + \underline{\underline{M_\gamma}}\,\underline{A} + \underline{\delta}\right) = \underline{\Delta} + \underline{\underline{M_\Gamma}}\,\underline{A} \tag{1.32}$$

where the matrices $[\underline{\underline{M_\alpha}}]_{ij} = \alpha_{ij}$, $[\underline{\underline{M_\beta}}]_{ij} = \beta_{ij}$, $[\underline{\underline{M_\gamma}}]_{ij} = \gamma_{ij}$ and $[\underline{\underline{M_\Gamma}}]_{ij} = \Gamma_{ij}$ are of dimension $m \times m$ and the vectors $[\underline{\delta}]_j = \delta_j$ and $[\underline{\Delta}]_j = \Delta_j$ are of dimension $m \times 1$. The corresponding matrix equation for ST can be obtained from (1.32) by simply replacing ϕ_j with $\hat{\phi}_j$. The set of equations (1.17c), (1.24) and (1.32) will be referred to as the GHF[m] model. The set of equations (1.17c), (1.24) and (1.32) with ϕ_j replaced with $\hat{\phi}_j$ will be referred to as the GST[m] model.

Obtaining LWE from the Galerkin approach.   We now demonstrate that LWE-HF and LWE-ST in § 1.3 can be obtained from an appropriate expansion of GHF and GST. For this purpose we assign to Re, Pe, We, M, B and B_w the same orders of magnitude as LWE. It is important to point out here that our averaged model in (1.32) has been derived without overly restrictive stipulations on the order of the dimensionless groups (see Kalliadasis et al. (2003a,b) for a discussion of lower/upper bounds on the order of magnitude of the dimensionless parameters). For example, changing the order of Pe in (1.32) would lead to a different long-wave expansion to that obtained in § 1.3.

Let us now expand q and the amplitudes θ and $A^{(i)}$ as $q = q^0 + \epsilon q^1 + O(\epsilon^2)$, $\theta = \theta_0 + \epsilon^{1-n}\theta_1$, $A^{(i)} = A_{0i} + \epsilon^{1-n}A_{1i}$ where $Pe = O(\epsilon^{-n})$ with $0 < n < 1$ and we truncate our expansions so that terms of $O(\epsilon^2)$ and higher in (1.24) are neglected while terms of $O(\epsilon)$ and higher in (1.32) are neglected. Equation (1.24) then yields

$$q = \frac{1}{3}h^3 + \frac{2}{5}Reh^6 h_x - \frac{1}{3}h^3 h_x \cot\beta - \frac{1}{2}Mh^2\theta_x + \frac{1}{3}Weh^3 h_{xxx}. \qquad (1.33)$$

We note that at this point θ_x from the averaged system in (1.32) remains undetermined, however, we shall demonstrate that it is exactly the same as the one obtained from LWE.

Substituting now q from equation (1.33) into equation (1.32), the ϵ^{1-n}-expansions for the temperature and utilizing the kinematic boundary condition (1.17c), yields

$$\theta = F + \frac{1}{40}PeBGh^3 h_x \ , \qquad A^{(2)} = \frac{1}{2}PeB^2(1 - B_w h)F^2 h^4 h_x$$

$$A^{(3)} = \frac{1}{8}PeBB_w(1 + 3Bh)F^2 h^4 h_x \ , \qquad A^{(4)} = -\frac{3}{40}PeB^2 B_w F^2 h^5 h_x.$$

We also have $A^{(i)} = 0$ for $i \geq 5$, for HF, while for ST we have

$$\theta = \hat{F} + \frac{1}{40}Pe\hat{B}\hat{G}h^3 h_x \ , \qquad A^{(2)} = \frac{1}{8}Pe\hat{B}(1 + 3\hat{B}h)\hat{F}^2 h^4 h_x \ , \qquad A^{(3)} = -\frac{3}{40}Pe\hat{B}^2\hat{F}^2 h^5 h_x$$

with $A^{(i)} = 0$ for $i \geq 4$. The expressions for θ are then substituted into (1.33), which in turn is substituted into the kinematic condition (1.17c) to yield LWE-HF given in equation (1.14) and the corresponding equation for LWE-ST. Further, when all the amplitude functions are substituted into equations (1.30a) and (1.30b), we obtain exactly the same temperature fields as those given in equations (1.12a) and (1.12b).

Hence, we have demonstrated that in order to obtain the long-wave theory of § 1.3 to $O(\epsilon Pe)$ from an appropriate expansion of our Galerkin system and hence fully resolve the behavior close to criticality we need $m \geq 4$ for HF and $m \geq 3$ for ST.

Single mode Galerkin approach. Although 4 and 3 are the minimum dimensionalities to fully resolve the behavior of small amplitude waves for HF and ST, respectively, for convenience we shall investigate the models obtained at the lower possible level of truncation, i.e. for $m = 1$. At this level the temperature profiles for HF and ST are quadratic and cubic, respectively, in η (and of course by construction they do satisfy all boundary conditions). The relative simplicity (at least compared to the higher-order projections) of the $m = 1$ models makes them useful prototypes for numerical and mathematical scrutiny. We then give explicitly (1.32) for $m = 1$. For GHF[1] we have:

$$\left[120 + 10\left(B^2 - 5BB_w + B_w^2\right)h^2 + 5\left(16 - BB_w h^2\right)\frac{h}{F} + \frac{6h^2}{F^2}\right]\frac{\theta_t}{30}$$

$$+ \left[80 + 35Bh + 55B_w h + (2B + 3B_w)^2 h^2 + (9 + 3Bh + 4B_w h)\frac{h}{F} + \frac{5h^2}{7F^2}\right]\frac{q\theta_x}{20h}$$

$$- \left(92B - 64B_w + 30B^2 h + 25BB_w h - 17B_w^2 h + \frac{15B + 4B_w}{F}h + \frac{27h}{7F^2}\right)\frac{\theta q_x}{120}$$

$$- \left[\frac{B^2 h + 4B - 2B_w}{2 + B_w h}\theta + \frac{2 + Bh}{2 + B_w h} + \frac{4}{21}(7 + B_w h) + \frac{11B}{168}(8 + B_w h)\right]\frac{2q_x}{5}$$

$$+ \left[324B - 96B_w - 84B_w^2 h - (56B_w + 43B)BB_w h^2 - \frac{16 + 74BB_w h^2}{F}\right]\frac{\theta q h_x}{280h}$$

$$+ \left[\frac{21}{F^2} + 12\frac{(B_w - B)^2}{2 + B_w h}\right]\frac{\theta q h_x}{70h} + \left(32 + 11Bh + 24\frac{2 + Bh}{2 + B_w h}\right)\frac{q h_x}{140}$$

$$+ \left[12 + 4(B + B_w)h + BB_w h^2\right]\frac{\theta F^{-1} - 1}{9 Peh} = 0 \tag{1.34a}$$

while for GST[1]:

$$\left(51 + 18\hat{B}h + 2\hat{B}^2 h^2\right)\frac{h\theta_t}{105} + \left(2137 + 698\hat{B}h + 73\hat{B}^2 h^2\right)\frac{q\theta_x}{3360}$$

$$+ \left[\left(\frac{101}{1344} - \frac{227}{3360}h\hat{B} - \frac{17}{960}\hat{B}^2 h^2\right)\theta - \frac{101}{1344} - \frac{19}{960}\hat{B}h\right]q_x + \left(349 + 73\hat{B}h\right)\frac{\hat{B}\theta q h_x}{3360}$$

$$+ (6 + \hat{B}h)\frac{\theta \hat{F}^{-1} - 1}{15 Peh} = 0. \tag{1.34b}$$

1.5 Neutral Stability Curves and Nonlinear Waves Far From Criticality

Table 1 summarizes the different models. We now examine the linear stability properties of our averaged models and in particular we compare the neutral stability curves of these models to those obtained from an Orr-Sommerfeld analysis of the full Navier-Stokes and energy equations (Trevelyan et al., 2006) for the HF and ST cases, referred to as OS-HF and OS-ST. The OS-ST problem has also been considered in detail by Scheid et al. (2005a). We also contrast the single-hump solitary wave solution branches obtained from all models. In all our computations we take $\beta = \frac{\pi}{2}$, $Ka = 3000$, $\hat{B}i = 0.1$ and in most cases $Pr = 7$, however, in order to assess the influence of the convective heat transport effects we shall some times include results with $Pr = 1$. The values for the parameters Ma, Bi_w, Bi and $\hat{M}a$ are chosen according to our analysis in §1.3 and so that LWE-HF and LWE-ST are identical as Pe tends to zero.

Neutral stability curves. Figure 2 illustrates typical curves for the neutral wavenumber for infinitesimal disturbances in the streamwise direction as a function of Re obtained from the averaged models and OS-HF/ST. The curves define the locus of vanishing real part of the growth rate of the disturbances.

Table 1. Summary of the equations for the different models.

LWE-HF/ST	SHF	SST	GHF[1]	GST[1]
(1.14)	(1.17c) (1.24) (1.29a)	(1.17c) (1.24) (1.29b)	(1.17c) (1.24) (1.34a)	(1.17c) (1.24) (1.34b)

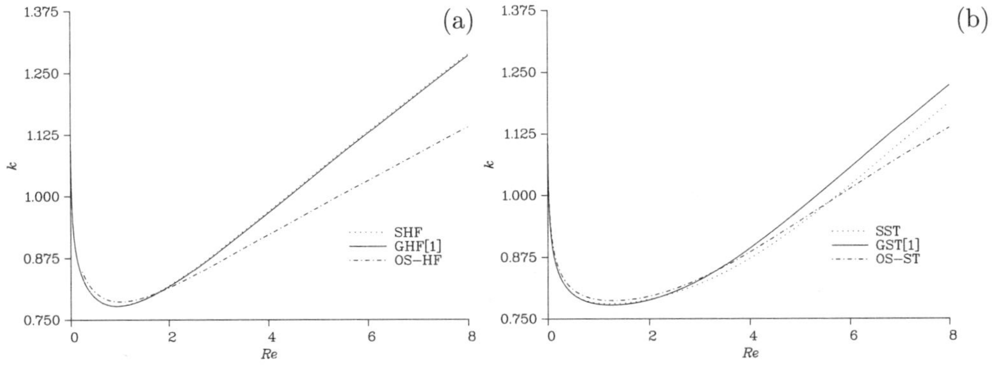

Figure 2. Neutral stability curves in the wavenumber k-Re plane for the different models with $Pr = 7$. In (a) (HF) $Ma = 21.6$, $Bi_w = 0.6$ and $Bi = 0.12$ and in (b) (ST) $\hat{Ma} = 30$ and $\hat{Bi} = 0.1$. OS-HF/ST denotes the Orr-Sommerfeld curves for the HF/ST problems, respectively.

As Re tends to zero the neutral wavenumber tends to infinity for both ST and HF cases. For the ST case this is in agreement with the OS-ST analysis by Scheid et al. (2005a). This behavior indicates that the Marangoni effect is stronger in the region of small film thicknesses. We shall return to this point when we discuss the nonlinear regime. As Re now increases the weighted residual models predict slightly smaller wavenumbers than OS-HF/ST initially, but further increasing Re shows that our models overpredict the neutral wavenumbers and they increasingly deviate from OS-HF/ST. This is to be expected as we have not taken into account the second-order dissipative effects which play an important role in the region of moderate to large Re. As was shown by Scheid et al. (2005a) for the ST case taking these terms into account leads to a good agreement with OS-ST for a much larger region of Reynolds numbers. Finally we note that for the HF case our models start to diverge from OS-HF at $Re \sim 2$ and they are almost graphically indistinguishable from each other over the entire range of Reynolds numbers in Figure 2(a). For the ST case both models follow a similar path, however the SST model performs better than the GST[1] with the divergence of SST at around $Re \sim 6$.

Solitary waves. We now seek traveling wave solutions propagating at speed c. We introduce the moving coordinate transformation $Z = X - c\Theta$ with $\partial/\partial\Theta = -c\partial/\partial Z$ for the waves to be stationary in the moving frame.

The resulting equation for LWE can then be integrated once with the integration constant determined from the far field condition $h(\pm\infty) = 1$ which leads to a nonlinear eigenvalue problem for the solitary waves speed c. For the weighted residuals models we also have the kinematic condition (1.17c) which in the moving frame yields $-ch' + q' = 0$. This can be integrated once and we fix the integration constant by demanding $(h, q) \to (1, \frac{1}{3})$ as $Z \to \pm\infty$. This gives a relation between the flow rate and the film thickness, $q = \frac{1}{3} + c(h - 1)$ which is then substituted into (1.24), (1.29a), (1.29b), (1.34a) and (1.34b) all in the moving coordinate system. The far field conditions $h(\pm\infty) = 1$ and $\theta(\pm\infty) = \overline{F}, \widehat{\overline{F}}$ then define the nonlinear eigenvalue problems for the solitary wave speed for the HF, ST cases respectively and the different models.

Here we shall restrict our attention to single-hump solitary waves, also called 'principal homoclinic orbits' by Glendinning and Sparrow (1984). We compute them using the continuation software AUTO97 with the HOMCONT option for tracing homoclinic orbits (Doedel et al., 1997). Figure 3 shows typical bifurcation diagrams for the speed c of the solitary waves as a function of Re. Interestingly as Re tends to zero the speed (and amplitude) of the solitary pulses tends to infinity. This is consistent with the linear stability analysis in Figure 2 which indicates that the influence of the Marangoni effect is larger for small Re.

This unusual behavior was first noticed for the ST case by Kalliadasis et al. (2003a) and was further discussed by Scheid et al. (2005a). As was pointed out by these authors in the limit of vanishing Reynolds number, inertia effects are negligible and the Marangoni effect is very strong. In this region of small film thicknesses, the destabilizing forces are interfacial forces due to the Marangoni effect (capillary forces are always stabilizing). For an isothermal falling film on the other hand, the only destabilizing forces are inertia forces which are vanishing as the Reynolds number tends to zero so that c in this region should approach the infinitesimal wave speed 1, as the $Ma = \widehat{Ma} = 0$ curves do. As was emphasized by Kalliadasis et al. (2003a) this behavior does not correspond to a true singularity formation as other forces of non-hydrodynamic origin (e.g. long-range attractive intermolecular interactions) which have not been included here would introduce a lower bound on the rate at which Re approaches zero, thus arresting the singularity formation. Nevertheless, the existence of large amplitude structures in the region of small Re indicates that the system would approach a series of 'drops' in time-dependent computations in this region (not done here).

Figures 3(a-c) indicate that both LWE models exhibit an unrealistic behavior with branch multiplicity and turning points at particular values of Re. By analogy with the isothermal case and as we discussed in the Introduction we expect that LWE exhibits a finite-time blow-up behavior for Re larger than those corresponding to the turning points. Obviously this catastrophic behavior is related to the non-existence of solitary waves and signals the inability of LWE to correctly describe nonlinear waves far from criticality. Moreover Figure 3(a) shows the existence of limit points where LWE simply terminates, not observed before in studies of the isothermal falling film. Our averaged models on the other hand predict the continuing existence of solitary waves for all Re, thus preventing the occurrence of non-physical blow-ups.

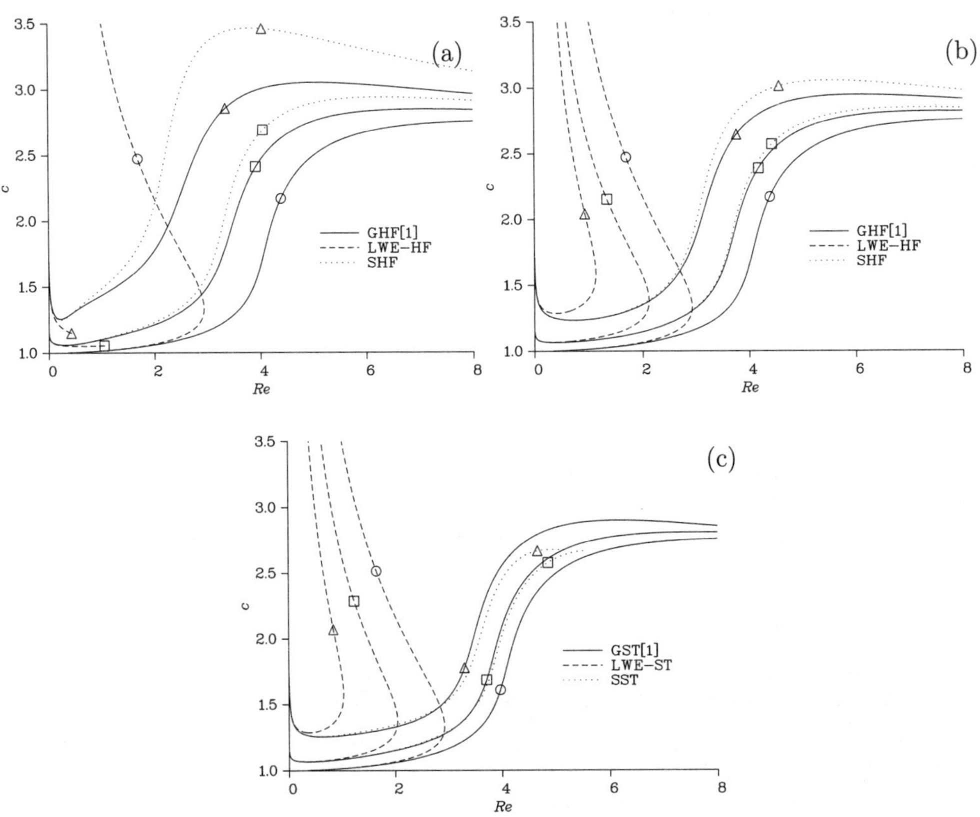

Figure 3. Single-hump solitary wave solution branches for the speed c as a function of the Reynolds number Re for the different models with $Pr = 7$. The circle corresponds to the isothermal case $Ma = \hat{Ma} = 0$. In (a) (HF) $Bi_w = Bi = 0.2$ with a square corresponding to $Ma = 12$ and a triangle to $Ma = 30$. In (b) (HF) $Bi_w = 0.6$ and $Bi = 0.12$ with a square corresponding to $Ma = 21.6$ and a triangle to $Ma = 54$. In (c) (ST) $\hat{Bi} = 0.1$ with a square corresponding to $\hat{Ma} = 30$ and a triangle to $\hat{Ma} = 75$.

For LWE we notice that the behavior in Figure 3(b) follows closely that in Figure 3(c) while in Figures 3(a-b) the SHF model predicts faster waves than the GHF[1]. On the other hand, in Figure 3(c) the SST model predicts slower waves than the GST[1]. Finally we note that all our weighted residuals models indicate that for sufficiently large Re the speed (and hence maximum amplitude) of the solitary pulses asymptote towards a certain value. As was pointed out for the ST case by Kalliadasis et al. (2003a), in this region of 'large' Re and hence 'large' film thicknesses, the interfacial Marangoni forces are not important compared to the dominant inertia effects.

1.6 Spatio-Temporal Dynamics: Evolution Towards Solitary Waves

We now illustrate the spatio-temporal dynamics of the heated falling film by using the GST[1] model (we found that GHF[1], SHF and SST behaved in a similar fashion). For our computations we employed a Crank-Nicolson-type implicit scheme with the spatial derivatives approximated by central differences and with dynamic time-step adjustment. We impose periodic boundary conditions over a domain much larger than the maximum growing wavenumber predicted by linear stability. The computations are performed in the moving frame $Z = (x - ct)/We^{1/3}$ with time given by $\Theta = t/We^{1/3}$.

Some typical time evolutions of the free surface and interfacial temperature are shown in Figure 4 for $\hat{Bi} = 0.1$, $\hat{Ma} = 30$ and $Pr = 7$. Figure 4(a) for $Re = 4$ shows that the final result of the evolution is a train of soliton-like coherent structures with almost the same amplitude and which interact indefinitely with each other like in soliton-soliton elastic collision. In figure 4(c) for $Re = 5$ the final result of the evolution is a single large amplitude wave preceded by a train of small-amplitude soliton-like coherent structures. The large wave collides with the smaller waves at the front and eventually absorbs them, leaving a flat region behind it. Due to the flat film instability, waves quickly begin to grow in this region involving towards a solitary wave train, however, due to the periodicity, they cannot escape from the large amplitude wave which collides with these smaller waves and overtakes them.

These coherent structures resemble the infinite-domain solitary pulses obtained in the previous sections. Interestingly in all cases the interfacial temperature is similar to an inverse of the free surface. Finally note that for the Re values in Figure 4 LWE does not have any solitary wave solutions (see Figure 3).

1.7 Finite-Differences Solution of the Energy Equation

We now contrast the interfacial temperature distribution obtained from our one-mode averaged models with the solution of the energy equation in (1.27), which is after all the equation we are trying to model. We set $Ma = \hat{Ma} = 0$. The reason for this is two-fold: (i) in this case the hydrodynamic and thermal problems are decoupled so that the temperature field does not have any influence on the film thickness; however, the evolution of the film thickness does affect the temperature field. We emphasize once again (see also our discussion in the Introduction) that Shkadov's IBL solitary wave solution branches are in quantitative agreement with the boundary-layer (Demekhin et al., 1987) and full Navier-Stokes equations (Demekhin and Kaplan, 1989; Salamon et al., 1993; Ramaswamy et al., 1996) while the single-mode corrected Shkadov model in (1.24) (with

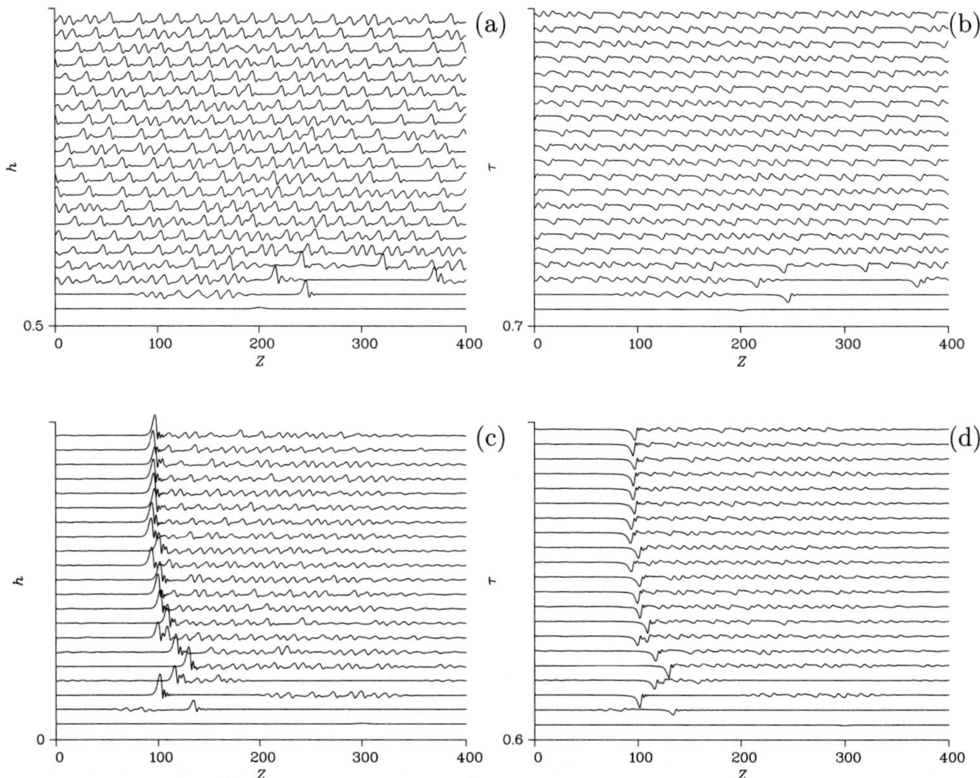

Figure 4. Time evolution for the free surface (a,c) and interfacial temperature (b,d) in an extended domain for GST[1] and in a coordinate system moving with speed c. $\hat{Bi} = 0.1$, $\hat{Ma} = 30$ and $Pr = 7$. In (a,b) $c = 1.1$. with $Re = 4$ with $c = 1.1$ and in (c,d) $c = 2.0$ with $Re = 5$. Successive curves are separated by (a,b) $\Delta\Theta = 100$ with $\Theta \in [0, 2 \times 10^3]$ and (c,d) $\Delta\Theta = 50$ with $\Theta \in [0, 1 \times 10^3]$.

$M = 0$) is in quantitative agreement with full Navier-Stokes (Ruyer-Quil and Manneville, 2000, 2002); (ii) the decoupling of hydrodynamics from energy significantly simplifies the computations of the temperature field.

Substituting the flow field from (1.21) and (1.5a) into (1.27), we obtain

$$T_{\eta\eta} = \frac{Pr\delta}{2} \left[\left(3qh(2\eta - \eta^2) - 2ch^2 \right) T_Z + chh_Z\eta(\eta - 1)(\eta - 2)T_\eta \right] \qquad (1.35a)$$

subject to the boundary conditions

$$T_\eta = -BhT \quad \text{on} \quad \eta = 1 \qquad (1.35b)$$

$$T_\eta = -h + B_w hT \quad \text{on} \quad \eta = 0 \quad \text{for HF} \qquad (1.35c)$$

$$T = 1 \quad \text{on} \quad \eta = 0 \quad \text{for ST} \qquad (1.35d)$$

which is solved in the (Z, η) domain along with periodic boundary conditions in the Z-direction and where h is obtained from the isothermal momentum equation (1.24) in the moving frame Z with $q = \frac{1}{3} + c(h-1)$. The above system was solved numerically using a finite-differencing scheme. We shall refer to the numerical solution of equations (1.35a), (1.35b) and (1.35c) as 'finite differences for HF' or FDHF and the numerical solution of equations (1.35a), (1.35b) and (1.35d) as 'finite differences for ST' or FDST.

We found that for small values of Re all models are in good agreement, as expected. For larger Re we observe a difference between the temperature distribution obtained from our weighted residuals models to that obtained from the finite differences solution. This difference increases as Re increases. Figure 5 compares the temperature distributions obtained for the different models for $Pr = 7$, $Ma = \hat{Ma} = 0$ and in the region of moderate Re . LWE is not included in the figure as in this region it does not predict the existence of solitary waves.

Figures 5(a,c,e) for $Re = 10/3$ show that the interfacial temperature distributions obtained from our averaged models for both HF and ST are quite close to the actual solutions obtained from FDHF and FDST, respectively. Figures 5(b,d,f) for $Re = 5$ show that the Galerkin models for both HF and ST predict a similar interfacial temperature minimum to that obtained from FDHF and FDST. The simplified models on the other hand do follow closely the front of the interfacial temperature wave but they overshoot the minimum.

Let us now compare HF and ST with each other. For this purpose we plot the normalized interfacial temperature distributions $\theta/\overline{F}$ for HF and $\theta/\hat{F}$ for ST so that both scalings lead to a flat film solution of unity. Figure 6 depicts the normalized FDHF and FDST solutions for $Re = 3$ and $Re = 4$.

At the front of the interfacial temperature wave both rescaled temperatures are very similar. There is, however, a difference for the temperature minimum, with the FDHF solution for $Bi = Bi_w = 0.2$ predicting the lowest minimum, and the FDST solution predicting the highest minimum. The FDHF model with $Bi = 0.12$ and $Bi_w = 0.6$ follows quite closely the FDST solution. The FDHF solution for $Bi = Bi_w = 0.2$ also predicts the longest tail for the rescaled interfacial temperature at the back of the wave. Further, an overshoot at the back of the wave is present for the FDHF solution for $Bi = Bi_w = 0.2$ unlike the other two solutions.

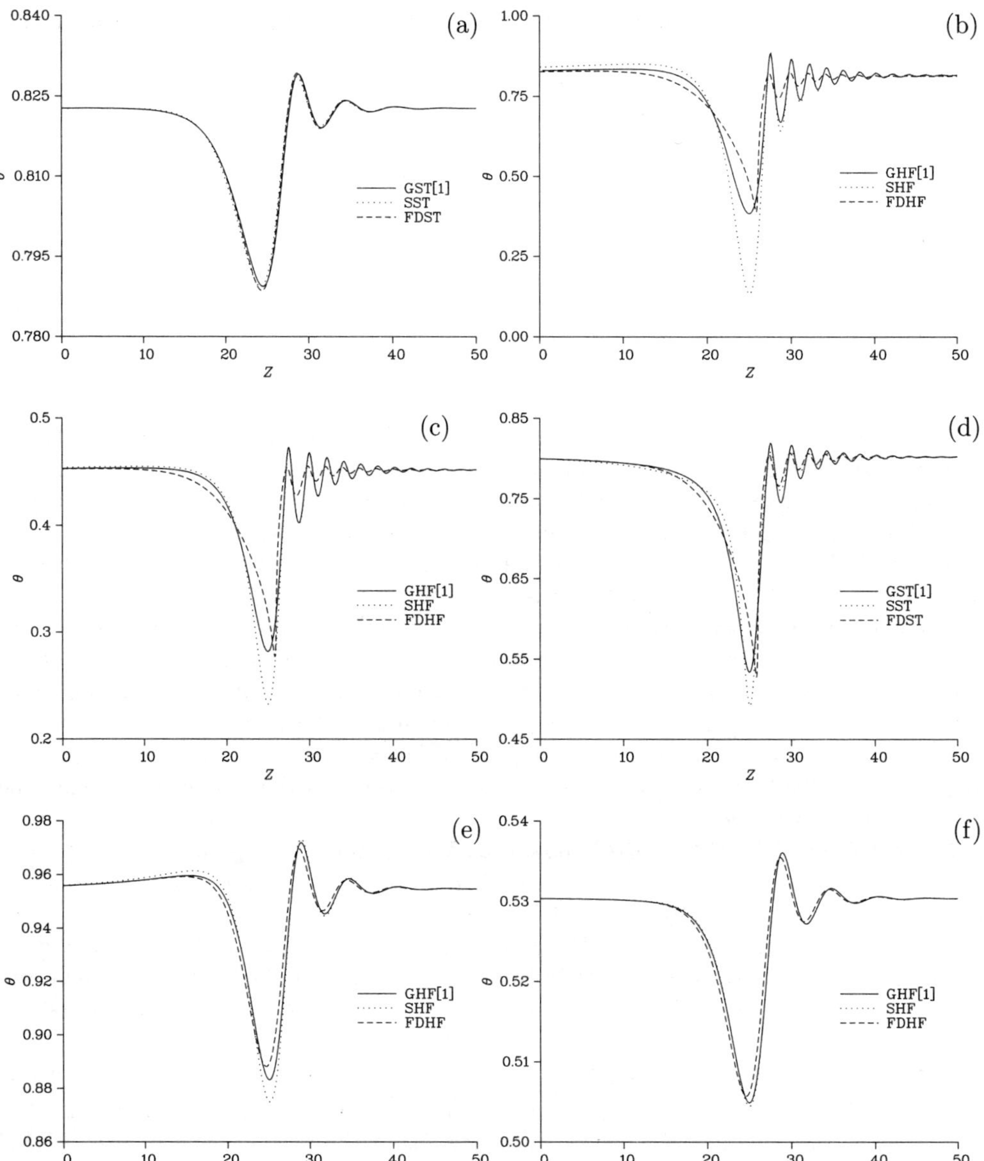

Figure 5. Comparison of the interfacial temperature distribution obtained from the different models for $Ma = \hat{Ma} = 0$ and $Pr = 7$. In (a, b) (HF) $Bi_w = Bi = 0.2$, in (c,d) (HF) $Bi_w = 0.6$ and $Bi = 0.12$ and in (e,f) (ST) $\hat{Bi} = 0.1$. In (a,c,e) $Re = 10/3$ and in (b,d,f) $Re = 5$.

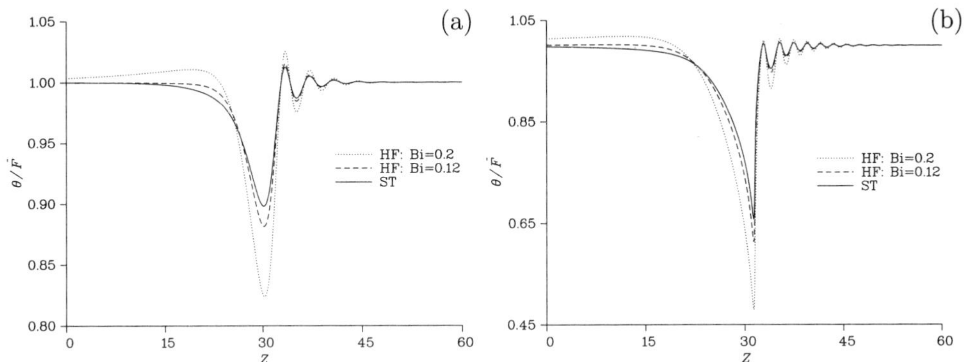

Figure 6. Normalized interfacial temperature for FDHF and FDST with $Ma = \hat{Ma} = 0$ and $Pr = 7$. In (a) $Re = 3$ and in (b) $Re = 4$. $Bi_w = 0.2$ when $Bi = 0.2$ and $Bi_w = 0.6$ when $Bi = 0.12$.

For the parameter values in the figure, we have qualitative similar solutions for the two cases, in fact FDHF for the second set of values, namely $Bi = 0.12$ and $Bi_w = 0.6$ which satisfies $Bi_w > Bi$ (see our discussion in § 1.3) produces results fairly close to those obtained from FDST. Hence, for convenience we shall only illustrate results for the ST problem in the remainder of this study.

With reference now to Figure 5, we have confirmed that all our averaged models are in good agreement with the finite-difference solutions provided that $Re < 4.5$. For larger values of Re a clockwise-turning recirculation zone appears inside the solitary wave. This zone is accompanied by two stagnation points, one at the front and one at the back of the wave. In this case solitary waves transport the trapped fluid mass downstream.

In Figure 7 we show streamlines and isotherms in the absence of the Marangoni effect with $Pr = 7$ and $Re = 4, 5$. As is evident from Figures 7(c,d) the presence of the recirculation zone alters dramatically the topology of the isotherms. For $Re = 4$ the isotherms are nearly aligned while for $Re = 5$ they are deflected upwards due to the movement of the fluid in the recirculation zone. This tightening of the isotherms occurs in the vicinity of the front stagnation point in the flow associated with the formation of steep temperature gradients there.

In Figure 8 we illustrate the effect of the Péclet number on the heat transport process before and after the recirculation zone appears. The interfacial temperature distribution is obtained from FDST. In Figure 8(a) we take $Re = 4$ in which case a recirculation zone does not exist while in Figure 8(b) we take $Re = 5$ in which case a recirculation zone is present. In the absence of a recirculation zone, increasing the Péclet number dampens the free-surface temperature distribution. On the other hand, when a recirculation zone is present, increasing the Péclet number causes a sharp gradient on the free-surface temperature distribution. Figure 8(c) shows that the sharp gradient appears in the vicinity of the front stagnation point corresponding to the formation of a thermal boundary layer and in fact the maximum of the temperature gradient moves towards the

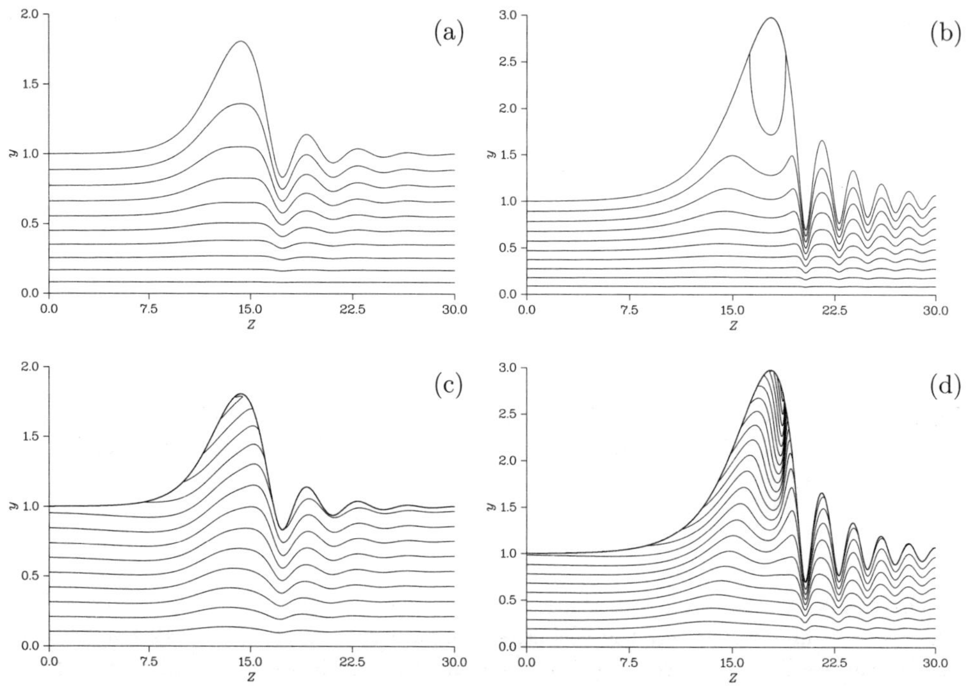

Figure 7. Streamlines (a,b) and isotherms (c,d) obtained from FDST with $Ma = \hat{Ma} = 0$, $\hat{Bi} = 0.1$ and $Pr = 7$. In (a,c) $Re = 4$ and in (b,d) $Re = 5$.

front stagnation point as Pr increases and is located exactly at this point in the limit of infinite Péclet number. Obviously, in this limit θ_x tends to infinity as well. We expect that in the presence of the Marangoni effect and for large Péclet numbers, the sharp spike in θ_x will have a significant influence on the fluid flow. It is quite likely that in this case thermocapillarity might cause the formation of a recirculation zone at smaller Re and might tighten both streamlines and isotherms due to enhancing the circulation in the primary solitary hump thus leading to both large temperature and velocity gradients in the flow.

Figure 9 compares the minimum of the interfacial temperature distribution $\theta_{\min}$ for ST as a function of Re obtained from SST and GST[1] to that obtained from FDST. Once the recirculation zone appears at $Re = 4.5$, FDST falls rapidly to a value of $\theta_{\min}$ 0.47 where it saturates as Re increases. The curve terminates at $Re = 7$ due to numerical difficulties with FDST at large Re.

Prior to the appearance of the recirculation zone both SST and GST[1] accurately model FDST. However, soon after the birth of the recirculation zone, the SST model appears to diverge away from FDST almost quadratically as Re increases. On the other hand, the GST[1] appears to follow FDST for a larger region of Reynolds numbers but

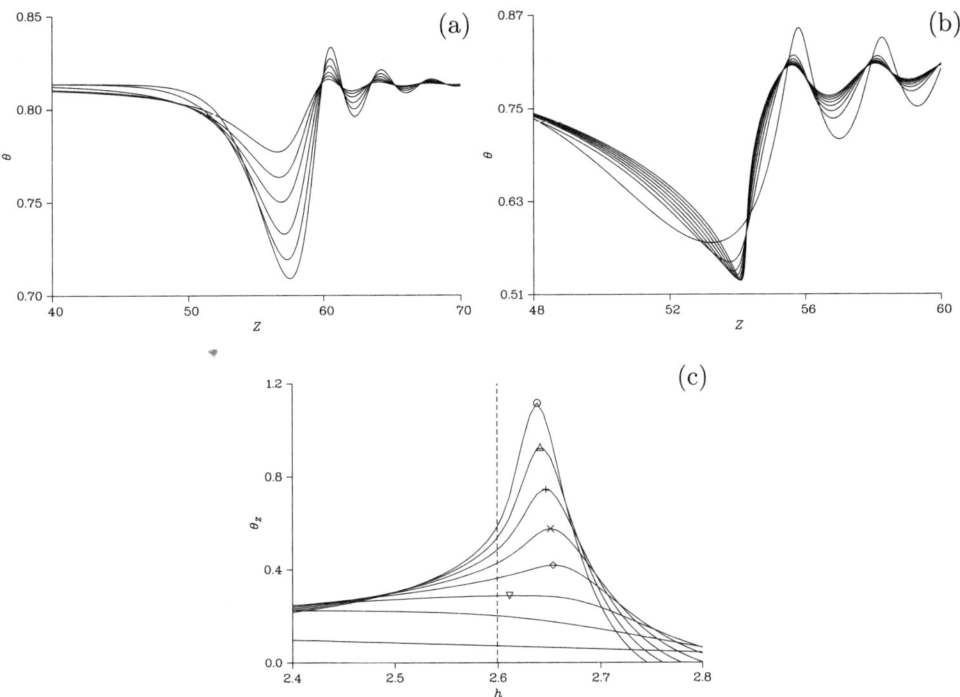

Figure 8. Interfacial temperature and crossplot of the interfacial temperature gradient as a function of h obtained from FDST with $Ma = \hat{Ma} = 0$ and $\hat{Bi} = 0.1$. In (a) $Re = 4$ and $Pr = 1, 3, 8, 20, 40$ and 90 and in (b, c) $Re = 5$ and $Pr = 0$ to 7 in increments of 1. The doted line in (c) denotes the location of the front stagnation point.

at some point it starts to diverge almost linearly thus predicting higher interfacial temperatures than SST and closer to the actual values.

Clearly the new model GST[1] shows a marked improvement over the previous SST model (recall from §1.4 that this model consists of (1.17c), (1.24) and (1.29b) which correspond to (5.17a), (5.17b) and (5.17c) in Scheid (2004)). Nevertheless, despite the improvement, at some point the new model also fails to describe accurately the interfacial temperature. In fact, both models give negative temperatures at some Re which are obviously unphysical: turning back to dimensional quantities, this would imply that the temperature on the fluid can be locally greater than the temperature of the wall or smaller than the air temperature, which is unrealistic.

The increased deviation between the interfacial temperature predicted by our models and the numerical solution of the energy equation is due to the formation of the boundary layer at the front stagnation point with large gradients in θ_Z. As a result, for large Re all models overshoot the minimum temperature to give negative temperatures which is a clear failure of the models. To accurately represent the temperature field as the boundary layer develops, one would need an increasingly large number of test functions. At some point and as the thickness of the boundary layer tends to zero the number of test functions should tend to infinity. Hence, any weighted residuals approach is bound to fail in this region. However, we do expect that physically it will be rather difficult for the system to sustain a two-dimensional thermal boundary layer and it is quite likely that three-dimensional effects and related instabilities (e.g. rivulet formation) will diffuse the sharp temperature peaks in the transverse direction.

Finally we note that both SST and GST[1] suffer from the coefficient of θ_Z passing through zero when the wave amplitude becomes sufficiently large. Consider for instance the Galerkin projection in (1.31) and (1.32). In the moving frame, the coefficient of θ_Z is $\langle (u^{(0)} - c)\phi_i\phi_j \rangle$. As the interfacial waves become larger, the flow becomes faster so that the term $u^{(0)} - c$ which is strictly negative for small amplitude waves, passes through zero and becomes positive (recall that the ϕ_i's are non-negative. Note that in the absence of Marangoni effects, $u^{(0)} = c$ on the interface when $h = 3 - (2/c)$. Hence, as the height of the waves increases, the inner product $\langle (u^{(0)} - c)\phi_i\phi_j \rangle$ can in fact change sign and become positive (recall that the ϕ_i's are non-negative).

Now $\theta_{\min} < 0$ and the coefficient of θ_Z going through zero, do not necessarily occur at the same time, in general $\theta_{\min} < 0$ happens first, nevertheless, while the models diverge from FDST and $\theta_{\min}$ decreases continuously towards zero, so does the magnitude of the coefficient of θ_Z. Eventually when this coefficient is zero, we encounter some serious difficulties in the numerical construction of the solitary waves. To alleviate these difficulties and examine the possibility that the divergence of the models from FDST and predictions of negative temperatures could be due to the reduction in magnitude of the coefficient of θ_Z as Re increases, we follow the suggestion by Trevelyan and Kalliadasis (2004c) and multiply the weight function for the energy equation with $u^{(0)} - c$ prior to averaging. This prevents the coefficient of θ_Z from going through zero.

This leads to the modified models MSST and MGST[1] for the ST case which never predict negative temperatures. Figure 9 shows that both MSST and MGST offer a substantial improvement over the SST and GST[1] models, nevertheless, a divergence (albeit slow) from FDST is observed and eventually these models also fail to represent the actual free-surface temperature distribution. Again this is due to the development of a thermal boundary layer at the front stagnation point. As we pointed out above, it is quite likely that the two-dimensional flow cannot sustain a sharp boundary layer and will attempt to diffuse it in the transverse direction with a development of a three-dimensional instability. Nevertheless, within the context of two-dimensional flows, it remains a challenge to obtain a model that accurately describes the interfacial temperature for the widest possible range of Re.

2 Three-Dimensional Wave Dynamics in Isothermal Thin Films

2.1 Introduction

As we discussed in § 1.1, a falling film is a convectively unstable open-flow hydrodynamic system that exhibits a rich variety of spatial and temporal structures and a rich

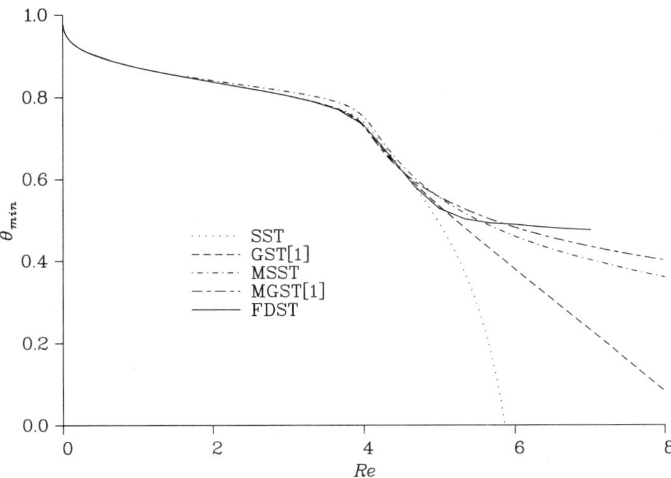

Figure 9. Minimum interfacial temperature $\theta_{\min}$ against Re with $\hat{Ma} = 0$, $\hat{Bi} = 0.1$ and $Pr = 7$ for ST.

spectrum of wave forms and wave transitions (many of which are generic to a large class of hydrodynamic systems). Intense research during the last few decades has elucidated the sequence of transitions of two-dimensional (2D) waves during their spatial evolution on the film (Chang and Demekhin, 2002). These transitions begin with a filtering mechanism of small amplitude white noise at the inlet that selects a monochromatic disturbance followed by a secondary modulation instability that transforms the primary wave field into a train of 2D soliton-like coherent structures (see also our discussion in § 1.1).

Assuming strong surface tension effects and long waves, a weakly nonlinear expansion of the 2D isothermal Navier-Stokes and free-surface boundary conditions yields the 2D Kuramoto-Sivashinsky (KS) equation,

$$\frac{\partial H}{\partial t} + 4H\frac{\partial H}{\partial x} + \frac{\partial^2 H}{\partial x^2} + \frac{\partial^4 H}{\partial x^4} = 0, \tag{2.1}$$

derived first by Shkadov (1973) (see also Nepomnyashchy (1974), Lin (1974) and Demekhin et al. (1983)). H, x and t are appropriately rescaled film thickness, streamwise coordinate and time, respectively. This equation is a well studied prototype because it retains (with the exception of dispersion) the fundamental elements of any nonlinear process that involves wave evolution in 2D: the simplest possible nonlinearity $H\partial H/\partial x$, instability and energy production $(\partial^2 H/\partial x^2)$, and stability and energy dissipation $(\partial^4 H/\partial x^4)$. Notice that the nonlinearity arises effectively from the nonlinear correction to the phase speed, a nonlinear kinematic effect that captures how larger waves move faster than smaller ones. For thin film flows, the nonlinearity is due to the interfacial kinematics associated with mean flow. At the same time, the functional form of the nonlinear term can be easily obtained from simple symmetry considerations: indeed the only other term that can be more dominant is H^2 which is obviously ruled out for systems whose spatial average does not drift, i.e. $d/dt\langle H\rangle_x = 0$.

With the addition of the dispersive term $\partial^3 H/\partial x^3$, the 2D KS equation becomes the 2D generalized KS (gKS) equation,

$$\frac{\partial H}{\partial t} + 4H\frac{\partial H}{\partial x} + \frac{\partial^2 H}{\partial x^2} + \delta\frac{\partial^3 H}{\partial x^3} + \frac{\partial^4 H}{\partial x^4} = 0, \tag{2.2}$$

where δ a positive parameter that characterizes the relative importance of dispersion and whose magnitude depends on the particular case considered. The 2D gKS equation has been reported for a wide variety of systems including a falling film with weak surface tension (Topper and Kawahara, 1978), a film falling down a uniformly heated wall (Kalliadasis et al., 2003a), plasma waves with dispersion due to finite ion banana width (Cohen et al., 1976) and liquid films sheared by a turbulent gas (Peng et al., 1991). Kawahara and Toh (1988) constructed numerically stationary solitary pulse solutions of the 2D gKS equation and showed that for large δ the pulse solutions become large in amplitude and fairly close to the symmetric sech^2-soliton shape for the 2D Korteweg-de Vries (KdV) equation, but, for smaller values of δ, an asymmetry develops in the pulse shape to yield an oscillatory structure at the front side tail of the pulse, while for much smaller values of δ this oscillatory structure of the front tail is enhanced. Note that the width of the pulses is almost the same for all δ as the coefficients of the second and fourth order derivatives in the 2D gKS equation are unity, which means fixing the wavenumber that gives the maximum linear growth rate. A detailed phase-plane analysis of all solitary wave solutions of the 2D gKS equation including multi-hump solitary waves has been performed by Nekorkin and Velarde (1994).

The time-dependent behavior of the 2D gKS equation has been scrutinized by Kawahara (1983) who demonstrated that the presence of $\partial^3 H/\partial x^3$ with a sufficiently large δ coefficient tends to arrest the spatial-temporal chaos exhibited by the 2D KS equation, in favor of a row of spatially periodic cellular structures each of which approaches the 2D KdV soliton as δ increases. The laminarizing effects of dispersion in the 2D gKS equation have also been considered by Chang et al. (1993) who constructed bifurcation diagrams for the periodic and solitary wave solutions of this equation, and also examined the linear stability of these solutions while Chang et al. (1995b, 1998) analyzed the response of solitary pulses to radiation wave packet disturbances.

Here we consider the three-dimensional (3D) gKS equation,

$$\frac{\partial H}{\partial t} - c\frac{\partial H}{\partial x} + 4H\frac{\partial H}{\partial x} + \frac{\partial^2 H}{\partial x^2} + \delta\frac{\partial}{\partial x}\nabla^2 H + \nabla^4 H = 0 \tag{2.3}$$

already transformed with $x \to x - ct$ to a coordinate system moving with constant speed c and where $\nabla^2 \equiv \partial^2/\partial x^2 + \partial^2/\partial z^2$. This equation is the simplest possible nonlinear evolution equation that retains the basic ingredients of nonlinear wave evolution in 3D. In the strongly dispersive limit, (2.3) describes a variety of physical phenomena that involve localized structures in 3D including solitary vortices in plasma (Zakharov and Kuznetsov, 1974), Rossby waves (Kuznetsov et al., 1986), magmons in magma segregation in the Earth's mantle (Scott and Stevenson, 1986) and localized rolls in nematic crystals (Joets and Ribota, 1988). In the context of thin liquid films, (2.3) has been derived for a film falling down a vertical plane assuming strong surface tension and near-critical conditions (Nepomnyashchy, 1974; Roskes, 1970; Lin and Krishna, 1977; Frenkel and Indireshkumar, 1999). In this case, the dispersion coefficient is found to be $\delta^2 = 3/(2ReWe)$ where Re and We are the Reynolds and Weber numbers defined from $Re = \bar{h}_N/3$ and $We = Ka/\bar{h}_N^2$ with $\bar{h}_N$ the dimensionless Nusselt flat film thickness and Ka the Kapitza number that depends on the fluid properties only (see §1.2). For $Ka \to \infty$ and fixed Re, $\delta \to 0$, however, for Ka large but fixed and $Re \to 0$, $\delta \to \infty$. These two limits have very different behavior due to the role of dispersion. Interestingly, the extra 3D dispersion term can be important even at large surface tension. This is distinct from the two-dimensional KS limit at large surface tension (see Chang and Demekhin (2002) for a detailed discussion of the regions in the parameter space where the different weakly nonlinear expansions apply).

In this study we examine 3D stationary solitary pulse solutions of (2.3). We obtain an analytical estimate for the speed c of these pulses in the strongly dispersive case, $\delta \gg 1$, by utilizing a perturbation from the 3D KdV equation. We also construct numerically 3D solitary pulses of (2.3) as a function of δ. For $\delta \to \infty$ our pulses approach the KdV limit with perfectly symmetric 3D pulses and a speed $c(\delta)$ obtained theoretically.

We note that 3D stationary solitary pulses were first obtained numerically by Petviashvili and Tsvelodub (1978) using the 3D KS equation. This study also introduced the term 'horse-shoe soliton' due to the fact that the shape of these waves strongly resembles a horse-shoe. As far as we are aware, since Petviashvili and Tsvelodub (1978) there has not been any progress in obtaining 3D solitons for equations of higher degree of complexity than KS.

Such 3D localized coherent structures have been observed experimentally on a falling film (Alekseenko et al., 1994; Tailby and Portalski, 1960; Park and Nosoko, 2003): after several spatial transitions of the naturally excited unstable disturbances, the initially 2D wave evolution on the falling film comes up to its last phase: at sufficiently large Reynolds numbers (of the order of 40 or so) the 2D solitary pulses become unstable to 3D perturbations and they eventually disintegrate into 3D coherent structures which are stable and robust and interact indefinitely with each other as 'quasi-particles' (but the bulk flow is still laminar). This stage of the evolution is often referred to as 'interfacial turbulence' or 'soliton-gas' Alekseenko et al. (1994). Interfacial turbulence is low-dimensional turbulence and persists up to a Reynolds number of the order of 1000-2000 (Chu and Dukler, 1974). Beyond this region we have the Tollmien-Schlichting instability and usual turbulence.

Of course, as pointed out earlier, the gKS equation is valid for near-critical conditions and therefore the Reynolds number must be small. Hence, an accurate description of interfacial turbulence in the context of the falling film problem would require a higher level of approximation than the weakly nonlinear expansion leading to (2.3). This is the IBL approximation discussed in §1.1 which in the region of large Reynolds numbers (up to 500 or so) is the model equation of choice (of course short of the complete Navier-Stokes). Quite recently we have been able to obtain particles of interfacial turbulence, i.e. localized stationary 3D solitary pulses for finite Reynolds number by using IBL. We have also been able to obtain with the same approximation the spatial evolution of 3D inlet disturbances from naturally excited room perturbations all the way up to 3D complex spatial-temporal dynamics. Our findings will be reported in a future study but a comparison with the 3D gKS equation indicates, that although this equation is valid for near-critical conditions and small amplitude waves, it can describe the basic characteristics of interfacial turbulence since, as we pointed out already, it retains the fundamental elements of any nonlinear process that involves wave evolution in 3D. Hence, we adopt the 3D gKS equation as a model system to describe interfacial turbulence in the context of thin films.

At the same time the 3D gKS can be viewed as a simple prototype to study the pattern formation dynamics and spatio-temporal complexity in active dispersive-dissipative nonlinear media. The time-dependent computations performed by Toh et al. (1989) with the 3D gKS equation indicate that for sufficiently large δ, the system involves into an aggregation of 3D solitary pulses so that its dynamics can be approximated in terms of these pulses. Hence, 3D solitary pulses become elementary processes representing the behavior of the full system so that its dynamics can be described by a superposition of these solitary pulses. The same is true for the falling film problem where despite the apparent complexity of the system, one can still identify 3D solitary pulses in what appears to be a randomly disturbed surface.

The idea that a complex wave pattern can be conveniently represented by interacting, coherent structures allows for a 'multiparticle' description to obtain equations of motion for the interacting pulses. This technique has been used in particle physics and condensed matter theory to describe particle interaction. Here we extend the inelastic coherent structure theory developed by Kawahara and coworkers Kawahara (1983); Kawahara and Toh (1988), Elphick et al. (1991, 1998), Balmforth et al., Chang et al. (1995a) and Chang and Demekhin (2002) among others to 3D pulses by including the additional degree of freedom due to the transverse direction. Notice that all these pulse interaction studies dealt with 2D waves only.

The main goal is to explain theoretically the V-shape formation of 3D waves observed in the time-dependent numerical experiments by Indireshkumar and Frenkel (1997). Our theory shows that, indeed, 3D solitary pulses for sufficiently large dispersion organize themselves into V-shapes. Our theoretical findings are in excellent agreement with our time-dependent computations of the fully nonlinear system.

2.2 Stationary Solitary Pulses

An analytical estimate of the speed of 3D solitary pulses for $\delta \gg 1$ is possible with a perturbation from the KdV soliton. For waves stationary in the moving frame, the transformation $H = \delta h$, $c = \lambda \delta$ and $\epsilon = 1/\delta$ where λ an $O(1)$ parameter yet to be determined, converts (2.3) to

$$
-\lambda \frac{\partial h}{\partial x} + 4h \frac{\partial h}{\partial x} + \frac{\partial}{\partial x} \left(\frac{\partial^2 h}{\partial x^2} + \frac{\partial^2 h}{\partial z^2} \right)
$$
$$
+ \epsilon \left(\frac{\partial^2 h}{\partial x^2} + \frac{\partial^2 h}{\partial z^2} + \frac{\partial^4 h}{\partial x^4} + 2 \frac{\partial^4 h}{\partial x^2 \partial z^2} + \frac{\partial^4 h}{\partial z^4} \right) = 0 \tag{2.4a}
$$

which by using the transformation $h = \lambda u$, $\partial/\partial x = \sqrt{\lambda}\partial/\partial\xi$, $\partial/\partial z = \sqrt{\lambda}\partial/\partial\eta$ is in turn converted to

$$\frac{\partial}{\partial\xi}\left(-u + 2u^2 + \frac{\partial^2 u}{\partial\xi^2} + \frac{\partial^2 u}{\partial\eta^2}\right) = -\epsilon\lambda^{-1/2}\left[\frac{\partial^2 u}{\partial\xi^2} - \lambda\left(\frac{\partial^4 u}{\partial\xi^4} + 2\frac{\partial^4 u}{\partial\eta^2\partial\xi^2} + \frac{\partial^4 u}{\partial\eta^4}\right)\right] \tag{2.4b}$$

which for $\epsilon = 0$ reduces to a 3D *free* of parameters KdV equation. Consider now the perturbation from the KdV soliton u^0, $u \sim u^0 + \epsilon\hat{u}$ which when substituted in (2.4b) gives

$$\mathcal{F}\hat{u} = -\lambda^{-1/2}\left[\frac{\partial^2 u_0}{\partial\xi^2} - \lambda\left(\frac{\partial^4 u_0}{\partial\xi^4} + 2\frac{\partial^4 u_0}{\partial\xi^2\partial\eta^2} + \frac{\partial^4 u_0}{\partial\eta^4}\right)\right] \tag{2.5a}$$

where $\mathcal{F}$ denotes the linear operator

$$\mathcal{F} = -\frac{\partial}{\partial\xi} + 4\frac{\partial}{\partial\xi}\left(u_0(\cdot)\right) + \frac{\partial^3}{\partial\xi^3} + \frac{\partial^3}{\partial\xi\partial\eta^2}. \tag{2.5b}$$

The adjoint of $\mathcal{F}$ with respect to the usual L^2 inner product, $\langle f, g\rangle = \int_{-\infty}^{+\infty}\int_{-\infty}^{+\infty} fg\,d\xi d\eta$, for any two functions f and g decaying sufficiently fast at the infinities, is found to be

$$\mathcal{F}^* = \frac{\partial}{\partial\xi} - 4u_0\frac{\partial}{\partial\xi} - \frac{\partial^3}{\partial\xi^3} - \frac{\partial^3}{\partial\xi\partial\eta^2}. \tag{2.6a}$$

But $\mathcal{F}^* u^0 \equiv 0$ is the equation defining u^0 since

$$\frac{\partial}{\partial\xi}\left(-u_0 + 2u_0^2 + \frac{\partial^2 u_0}{\partial\xi^2} + \frac{\partial^2 u_0}{\partial\eta^2}\right) = 0. \tag{2.6b}$$

Hence, u^0 is a null eigenfunction of the adjoint problem. To invert now the singular operator $\mathcal{F}$ and solve for $\hat{u}$, the right-hand-side of (2.5a) must satisfy the Fredhölm alternative condition:

$$\int_{-\infty}^{+\infty}\int_{-\infty}^{+\infty} u_0\frac{\partial^2 u_0}{\partial\xi^2}d\xi d\eta - \lambda\int_{-\infty}^{+\infty}\int_{-\infty}^{+\infty} u_0\left(\frac{\partial^4 u_0}{\partial\xi^4} + 2\frac{\partial^4 u_0}{\partial\eta^2\partial\xi^2} + \frac{\partial^4 u_0}{\partial\eta^4}\right)d\xi d\eta = 0. \tag{2.7}$$

Let us now introduce in this condition polar coordinates (ρ, θ) with $\rho = \sqrt{\xi^2 + \eta^2}$ and $\theta = \arctan(\xi/\eta)$. This allows to have ρ as the *only* independent variable

$$\int_0^\infty\int_{-\pi}^\pi \left(\frac{\partial u_0}{\partial\rho}\right)^2 (\cos\theta)^2 d\theta d\rho = \lambda\int_0^\infty\int_{-\pi}^\pi \frac{1}{\rho}\frac{\partial}{\partial\rho}\left[\rho^2\left(\frac{\partial u_0}{\partial\rho}\right)^2\right]d\theta d\rho \tag{2.8a}$$

or

$$\lambda = \frac{1}{2}\frac{\int_0^\infty(\frac{\partial}{\partial\rho}u_0)^2 d\rho}{\int_0^\infty \frac{1}{\rho}\frac{\partial}{\partial\rho}[\rho^2(\frac{\partial}{\partial\rho}u_0)^2)]d\rho} \tag{2.8b}$$

where u_0 satisfies the free of parameters KdV equation in polar coordinates

$$\frac{1}{\rho}\frac{\partial}{\partial\rho}\rho\left(\frac{\partial}{\partial\rho}u_0\right) - u_0 + 2u_0^2 = 0, \tag{2.9}$$

which can be easily solved numerically. A single numerical integration then gives $u^0(\rho)$ which when substituted in (2.8b) yields, $\lambda \simeq 0.3256$ or

$$c(\delta) \sim 0.3256\delta, \qquad \delta \gg 1. \tag{2.10}$$

Note that this velocity is about three times smaller than the estimate $c \sim (7/5)\delta$ for $\delta \gg 1$ of the corresponding 2D waves of the gKS equation obtained by Kawahara and coworkers (Kawahara and Toh, 1988; Kawahara, 1983). Hence, 3D localized coherent structures are much slower than 2D ones. At the same time the amplitude of 3D waves is also smaller than that of 2D's (roughly two times smaller). This observation has significant consequences for heat/mass transport in a falling film: for heat/mass transport enhancement we must maintain the 2D wave regime. However, issues related to flow control (by e.g. changing inlet conditions) in order to obtain optimal wave regimes for heat/mass transport are beyond the scope of the present study.

The full treatment for large δ including an approximate analytical solution for the shape of the 3D solitary pulses of (2.3) is given by Saprykin et al. (2005). Here we construct numerically 3D waves from the stationary version of (2.3) using continuation from the $\delta = 0$ limit (the numerical solution of the KS equation is easier than that of the gKS equation) and imposing periodic boundary conditions in an extended domain. The analytical estimate for $c(\delta)$ given above, can then be used to check the accuracy of our numerical scheme. Figure 10 depicts the quantity $\lambda = c/\delta$ as a function of δ. The analytical estimate is approached rapidly by the numerical solution and for $\delta \gtrsim 50$ the two are practically indistinguishable.

Figure 11 depicts 3D stationary pulse solutions of (2.3)). They are in qualitative agreement with the recent experiments by Alekseenko et al. (2005) for vertical falling films with $Re \sim 1$. 3D stationary pulses are typically characterized by a large curvature 'head' with capillary ripples at its front and two oblique 'legs' extending to the back of the head. Behind the head and between the two legs, there is a well pronounced dent caused by capillary depression in the legs. The depth of this dent varies slowly with z (it appears to be nearly constant for the domain size in Figure 10) and eventually vanishes at very large distances from the primary solitary hump. As δ increases, the pulse amplitude increases, the front running capillary waves decrease, the dent becomes less pronounced and the waves become increasingly symmetric. Eventually, as $\delta \to \infty$ we approach the KdV limit with perfectly symmetric 3D pulses and a speed $c(\delta)$ obtained above.

2.3 Coherent Structure Theory

We now formulate a coherent structure theory to describe the interaction of 3D horse-shoe waves. We begin by considering the interaction between two stationary pulses. Let us denote the two pulses as $H_1(x - X_1, z - Z_1)$ and $H_2(x - X_2, z - Z_2)$, located at (X_1, Z_1) and (X_2, Z_2), respectively, with X_i, Z_i slow functions of time (see Figure 12). The pulse $H_i(x - X_i, z - Z_i)$ is a shift of the $H(x, z)$ pulse to $x = X_i$ and $z = Z_i$. H_1 and H_2 are the same but at different locations so that for an infinite domain, $H_1 \equiv H_2$ is an exact solution. When the two pulses are brought together, they interact through their (exponentially decaying) tails since they overlap. We then introduce an overlap function

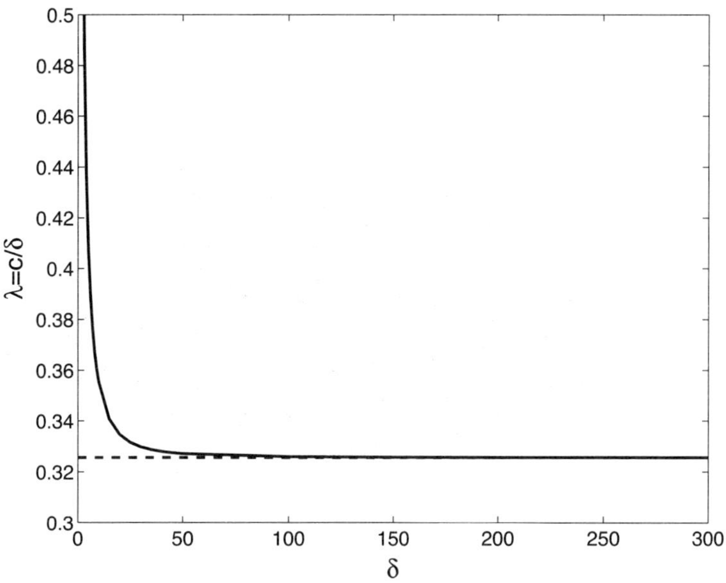

Figure 10. $\lambda = c/\delta$ as a function of δ. The dashed line is the asymptotic estimate, and the solid line is obtained from the fully nonlinear system.

$\hat{H}(x, z, t)$ and represent the profile as

$$H(x, z, t) = H_1(x - X_1, z - Z_1) + H_2(x - X_2, z - Z_2) + \hat{H}(x, z, t). \qquad (2.11)$$

The aim here is to represent complex wave patterns (e.g. interfacial turbulence on the falling film) as a *weak* interaction of 3D coherent structures. In the context of the falling film, 'weak' means that there is no mass transfer from one coherent structure in the flow to another and solitons just repel or attract each other by interacting through their tails only.

Let us first examine the influence of H_2 on H_1. $\hat{H}$ is then a perturbation or correction function for the 1st pulse H_1. Substituting (2.11) into (2.3) and neglecting the nonlinear term $\hat{H}\hat{H}_x$ yields an equation for $\hat{H}$

$$\frac{\partial}{\partial t}\hat{H} - \mathcal{L}\hat{H} = -4\frac{\partial}{\partial x}H_1 H_2 \qquad (2.12a)$$

where the linear operator $\mathcal{L}$ is given by

$$\mathcal{L} = c\frac{\partial}{\partial x} - \frac{\partial^2}{\partial x^2} - \delta\frac{\partial}{\partial x}\nabla^2 - \nabla^4 - 4\frac{\partial}{\partial x}(H_1(\cdot)). \qquad (2.12b)$$

In this equation we are in a frame of reference moving with the speed c of both pulses in an infinite domain and we focus on the vicinity of the 1st pulse which is taken to

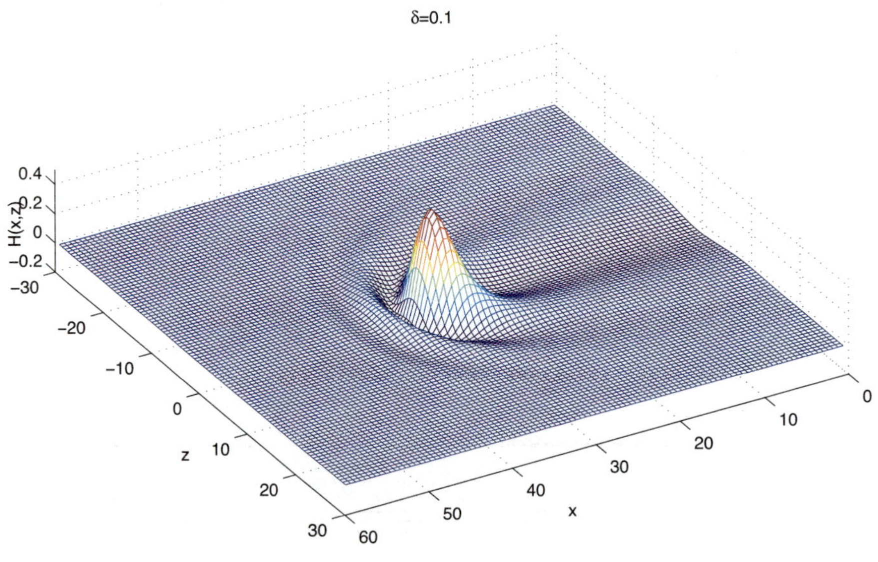

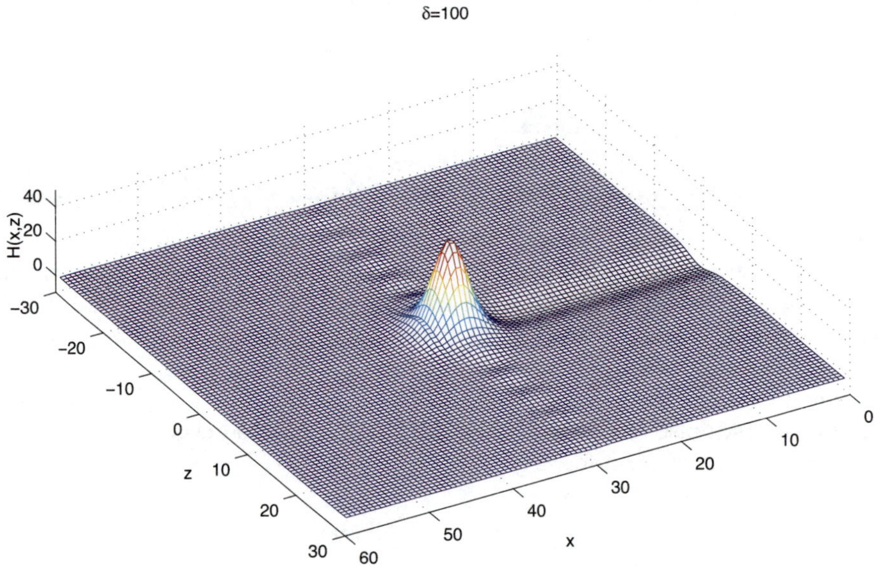

Figure 11. Stationary horse-shoe waves of (2.3).

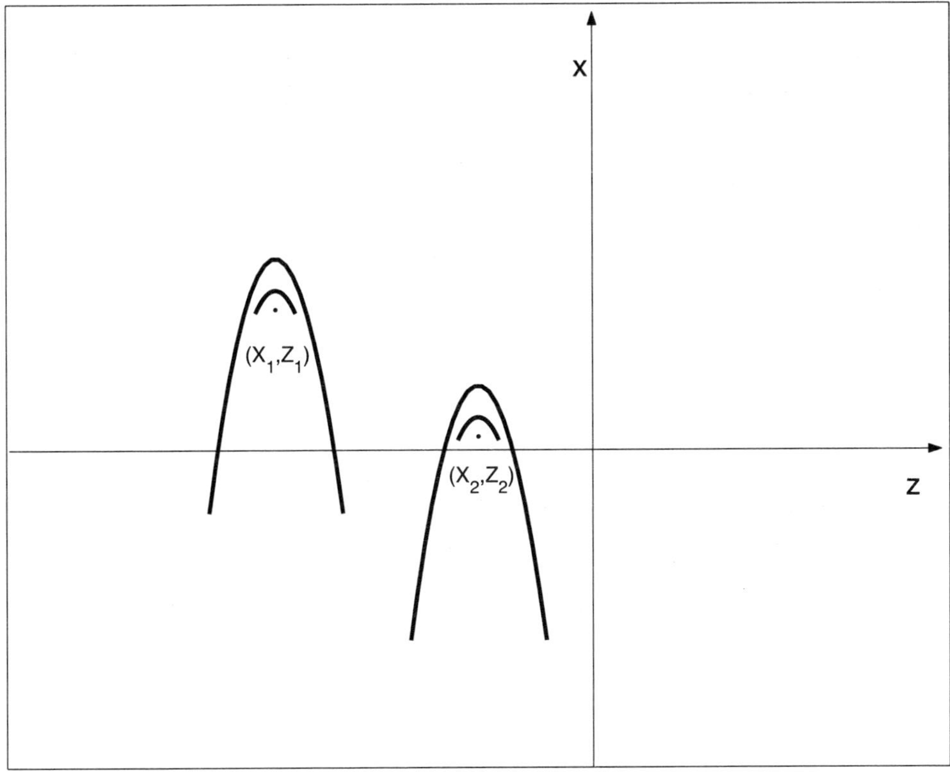

Figure 12. Schematic of a soliton pair. The two solitons are located at (X_1, Z_1) and (X_2, Z_2), respectively.

be stationary in this moving frame. Hence, $(H_2\hat{H})_x \ll (H_1\hat{H})_x$ since H_2 is small in the vicinity of the 1st pulse, i.e. far from the 2nd solitary hump where H_2 is 'large'. At the same time $(H_2\hat{H})_x \ll (H_1H_2)_x$ since $\hat{H}$ is small. Hence, the term $(H_2\hat{H})_x$ in (2.11) is neglected as both H_2 and $\hat{H}$ are small (unlike e.g. $(H_1H_2)_x$ where only H_2 is small). This is a key point of our theory. By neglecting this term, $\mathcal{L}$ depends on H_1 only. As a consequence, $\mathcal{L}$ has a two-dimensional null space spanned by the eigenfunctions $\phi_1 = H_{1x}$ and $\phi_2 = H_{1z}$ associated with the translational invariance of the system in the x- and z-directions, respectively. This also implies that the interaction of the two pulses is effectively due to translation only which in turn is due to the fact that the two solitons are sufficiently far from each other so that the influence of H_2 on H_1 is *linear* and can be described by (2.12a).

We now assume that the higher order eigenvalues of $\mathcal{L}$ have sufficiently negative real parts so that the higher order modes decay sufficiently fast. We then project the dynamics onto the null space, $\hat{H} \sim \mathcal{A}_1(t)\phi_1(x,z) + \mathcal{A}_2(t)\phi_2(x,z)$. The amplitudes $\mathcal{A}_{1,2}$ of the projection can be identified with the location of the second solitary pulse. Indeed, if we perturb a soliton by translating it slightly, $H(x-X(t), z-Z(t)) \sim H(x,z) - XH_x - ZH_z$ where $-XH_x - ZH_z$ is the *perturbation due to translation*. Hence, $\mathcal{A}_1 \equiv X$ and $\mathcal{A}_2 \equiv Z$.

Before we proceed further we also need to obtain the two null adjoint eigenfunctions which satisfy $\mathcal{L}^*\psi = 0$ where the null adjoint operator $\mathcal{L}^*$ with respect to the L^2 inner product defined above is found to be

$$\mathcal{L}^* = -c\frac{\partial}{\partial x} - \frac{\partial^2}{\partial x^2} + \delta\frac{\partial}{\partial x}\nabla^2 - \nabla^4 + 4H_1\frac{\partial}{\partial x}.$$

The adjoint eigenvalue problem was solved numerically, with boundary conditions decaying functions at the infinities, to obtain the two eigenfunctions $\psi_{1,2}$. We then substitute the projection for $\hat{H}$ into (2.12a) and by taking the inner product of both sides of the resulting equation with $\psi_{1,2}$ yields two amplitude equations for the location of the 2nd pulse,

$$\langle\phi_1,\psi_1\rangle\frac{\partial}{\partial t}X_2 = -4\langle\frac{\partial}{\partial x}H_1H_2, \psi_1(H_1)\rangle \tag{2.13}$$

$$\langle\phi_2,\psi_2\rangle\frac{\partial}{\partial t}Z_2 = -4\langle\frac{\partial}{\partial x}H_1H_2, \psi_2(H_1)\rangle \tag{2.14}$$

where the notation $\psi_{1,2}(H_1)$ implies that $\psi_{1,2}$ depend on H_1 only since $\mathcal{L}^*$ depends on H_1 only as well. These two amplitude equations describe the influence of H_2 onto H_1 (recall that H_2 is the moving pulse). Similarly, the influence of H_1 onto H_2 is described by the amplitude equations

$$\langle\phi_1,\psi_1\rangle\frac{\partial}{\partial t}X_1 = -4\langle\frac{\partial}{\partial x}H_1H_2, \psi_1(H_2)\rangle \tag{2.15}$$

$$\langle\phi_2,\psi_2\rangle\frac{\partial}{\partial t}Z_1 = -4\langle\frac{\partial}{\partial x}H_1H_2, \psi_2(H_2)\rangle. \tag{2.16}$$

Let us now define a distance vector between the two pulses, $\underline{L} = (L_x, L_z)^t$ where $L_x = X_2 - X_1$ and $L_z = Z_2 - Z_1$. By subtracting (2.15,2.16) from (2.13,2.14), we obtain a dynamical system for $\partial \underline{L}/\partial t$:

$$\langle \phi_1, \psi_1 \rangle \frac{\partial}{\partial t} L_x = -4 \langle \frac{\partial}{\partial x} H_1 H_2, \psi_1(H_1) \rangle + 4 \langle \frac{\partial}{\partial x} H_1 H_2, \psi_1(H_2) \rangle \qquad (2.17)$$

$$\langle \phi_2, \psi_2 \rangle \frac{\partial}{\partial t} L_z = -4 \langle \frac{\partial}{\partial x} H_1 H_2, \psi_2(H_1) \rangle + 4 \langle \frac{\partial}{\partial x} H_1 H_2, \psi_2(H_2) \rangle. \qquad (2.18)$$

We can then obtain the fixed points of this system and determine the basins of attraction. Figure 13 shows the interaction map obtained from (2.17, 2.18) for $\delta = 100$. There are countably infinite fixed points that arise in (2.17, 2.18). These are alternately saddle points and stable nodes. Here we only show a few of the fixed points in the immediate vicinity of the primary solitary pulse at $(0,0)$. At large distances from the primary soliton, the interaction between the pulses is fairly weak and it becomes increasingly difficult to locate the fixed points.

We have also obtained numerically the interaction map from the fully nonlinear problem in (2.3). The numerical scheme employs a global Fourier spectral expansion in both coordinates x and z. The integration domain is sufficiently long for three pulses so that there is no interaction with their periodic extension. We were able to determine three fixed points: a saddle and two attractors. Note that both our theory and time-dependent computations indicate that the solitary pulses form a V-shape pattern. This self-organization is only possible for large values of δ. We have hence provided theoretical evidence for the V-shape formation of 3D horse-shoe waves observed in the time-dependent numerical experiments with (2.3) by Indireshkumar and Frenkel (1997). Note that this V-shape formation was not observed in the computations reported by Toh et al. (1989) due to the δ value used by these authors ($= 25$ while to observe the V-shapes δ must be $\gtrsim 50$). Figure 14 depicts the variation of stable and unstable equilibria as a function of δ. Notice the excellent agreement between the theory and the fully nonlinear computations for ϕ and L_x for the 1st stable point as a function of δ. The reason for this is two-fold: (i) the distance between the two trailing solitons is always larger than the distance between each of these solitons and the front soliton, e.g. for $2\phi \sim O(90^0)$, the distance between the trailing solitons is ~ 1.5 times the distance between each of the trailing solitons and the front one; (ii) the overlap between the legs of the leading soliton and the tails of the trailing solitons is much stronger than the overlap of the legs of the two trailing solitons. Interestingly, for $\delta \gg 1$, ϕ approaches an asymptote.

Figure 15 shows the formation of a V-shape at $t = 150$ obtained from the fully nonlinear system in (2.3) for $\delta = 100$. The initial condition is a 3D wave of amplitude smaller than that of a 3D stationary pulse solution for the same δ. After 110 time units the initial condition has disintegrated into three solitary pulses with the middle pulse larger than the two neighboring ones which have roughly the same amplitude. The final asymptotic state is a V-shape that consists of three solitary pulses of the same amplitude and speed. Careful examination of the trailing solitons indicates that they are located at the legs of the leading soliton. In fact, time-dependent experiments with a single 'excited' soliton, i.e. a pulse of larger amplitude to that corresponding to a given δ and formed by e.g. adding some mass to the primary solitary hump, show that the soliton will drain the additional mass to the back through the legs. This mass will then form two additional

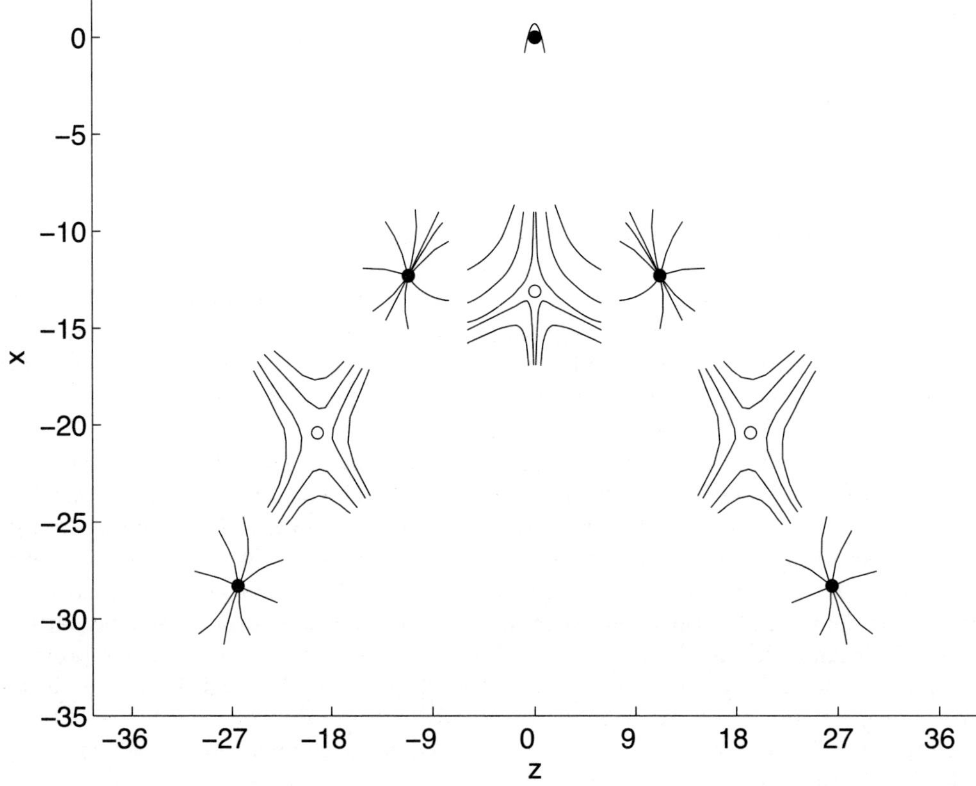

Figure 13. Interaction map obtained from (2.17, 2.18) for $\delta = 100$. The stationary pulse is located at $(0,0)$. Saddle points are denoted by $\circ$ and attractors by $\bullet$. The lines surrounding these points are the orbits of the 2nd soliton.

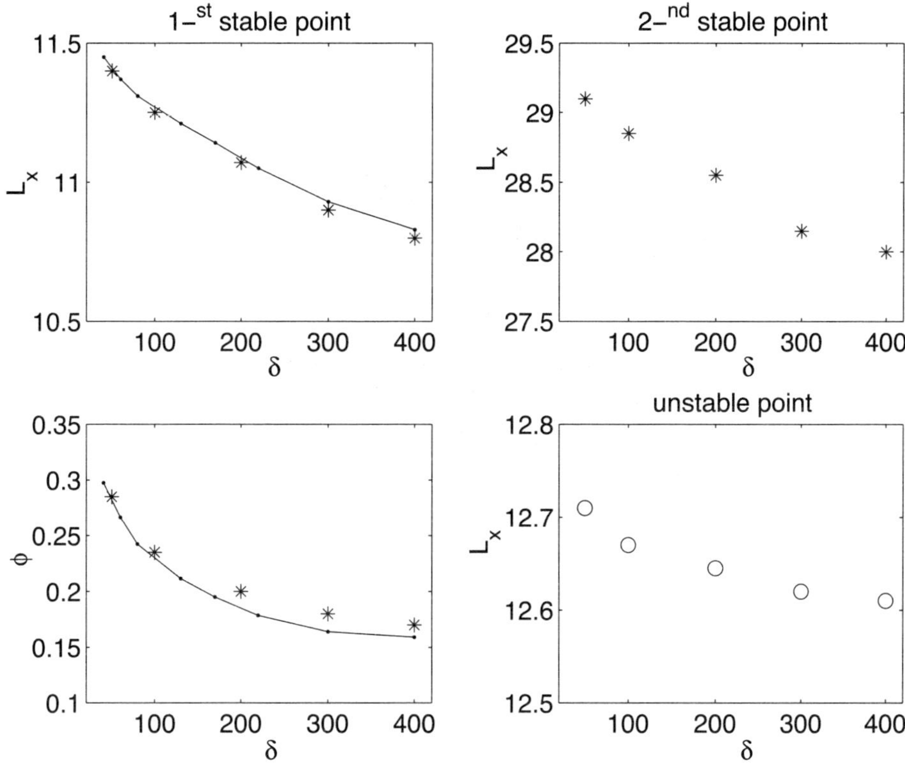

Figure 14. Variation of stable and unstable equilibria as a function of δ obtained from our interaction theory: '1st stable point' and 'unstable point' denote the attractor and saddle point, respectively, in the immediate vicinity of the stationary pulse. '2nd stable point' denotes the attractor located further than the 1st stable point. Also shown as a function of δ is the angle $\phi = (1/2)\times$ (V-shape angle) – the location of the fixed points is fully determined by ϕ and L_x. The continuous lines in the (δ, L_x) graph for the 1st stable point and (δ, ϕ) graph are obtained from the full system in (2.3).

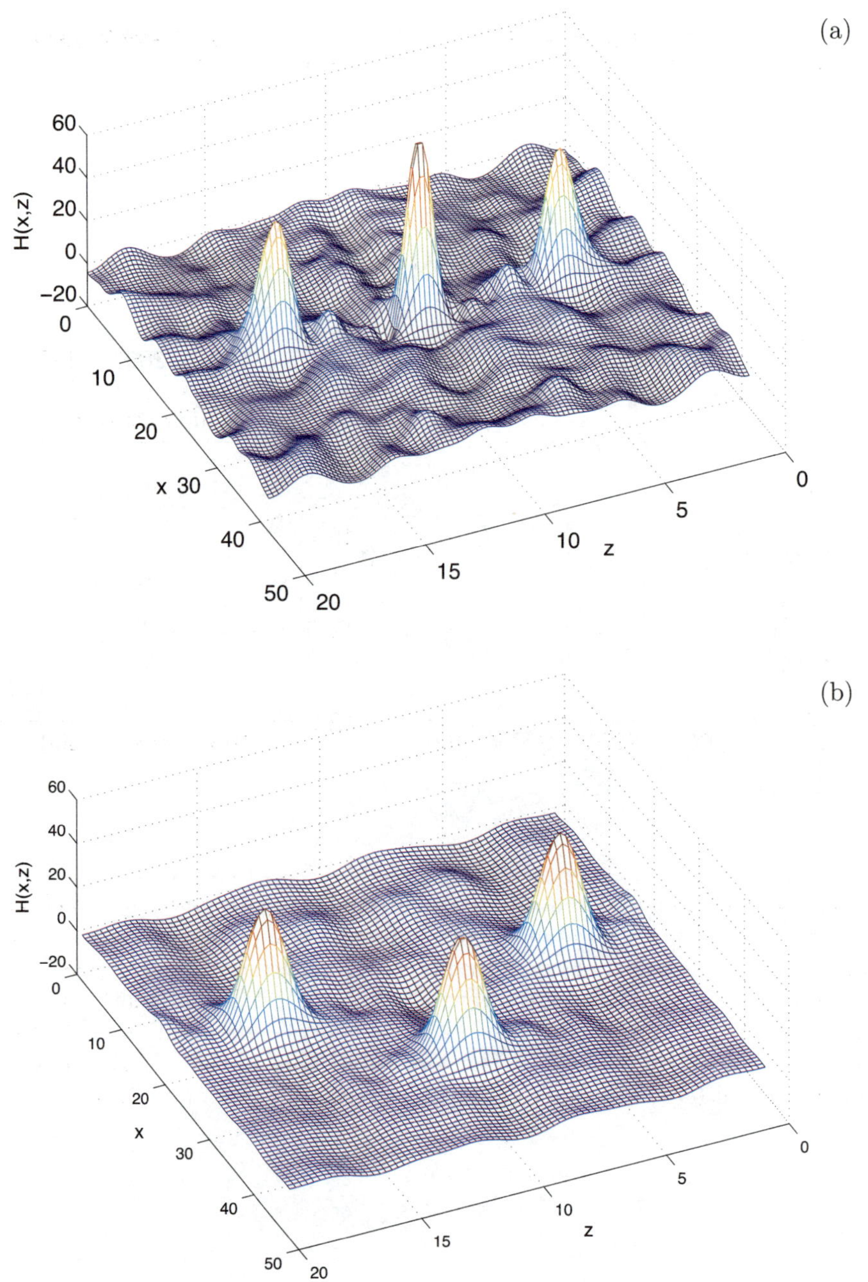

Figure 15. 3D time-dependent evolution for $\delta = 100$ using as an initial condition a 2D wave of amplitude smaller than that for a 2D stationary soliton for the same δ; (a) Snapshot at $t = 110$. The originally 2D wave has been transformed into three 3D solitons; (b) V-shape formation at $t = 150$. After 40 time units the three solitons in (a) self-organize into a V-shape.

pulses at the legs of the leading soliton. Note that the small-amplitude background noise surrounding the three solitons is due to the KS chaos which is still present at $\delta = 100$.

Figure 16 shows the formation of a V-shape with five solitary pulses at $t = 170$ for $\delta = 100$. The initial condition is again a 3D wave of amplitude smaller than that of a 3D stationary solitary pulse for the same δ. But now the domain is larger than in Figure 15 (truncated in Figure 16 for presentation purposes). Obviously, the number of solitons depends on the computational domain: the bigger the domain, the larger the number of waves (for a given δ). If the number of waves is sufficiently small, e.g. three or five, and the domain sufficiently large, then the waves will organize themselves into a single V-shape of the type shown in Figures 15 and 16. If we have many pulses and a very large domain then there are many different possibilities. Two such possibilities are depicted in Figure 17. At the top of the figure we have a smaller V-shape within the main V-shape formed by two of the solitons on the right branch of the main V-shape and a third soliton located between the two branches of the main V-shape. In this case the angle of the smaller V-shape is roughly the same as that of the main V-shape which in turn is fairly close to that predicted by our theory.

Another possibility is shown at the bottom of Figure 17. We now have two separate V-shapes, each with three solitons, of the type observed in Figure 15. This configuration was observed for $\delta = 100$ in which case the distance between two neighboring solitons is $\simeq 15$. Provided that the two V-shapes never come close within a distance d smaller than $\simeq 30$ (i.e. roughly twice the distance between two neighboring pulses in a V-shape), the angles of the two V-shapes are roughly the same and fairly close to what is predicted by our theory. For d smaller than about 30, the two V-shapes begin to interact in a complex fashion which alters their angles to values which cannot be predicted by our binary pulse interaction theory.

In this case we would have to consider the many-body problem. We have already demonstrated the basic concept for such a multiparticle description by investigating the interaction between two 3D equilibrium pulses. For three-pulse interaction, say $H_{1,2,3}$ with H_1 the primary soliton, we must neglect the terms $(\partial/\partial x)(H_2 \hat{H})$ and $(\partial/\partial x)(H_3 \hat{H})$, assuming that particles 2 and 3 are now located far from the primary particle 1. The right-hand-side of (2.12a) is then replaced by $-4(\partial/\partial x)(H_1 H_2 + H_1 H_3 + H_2 H_3)$. Hence, unlike 2D pulse interaction, where only neighoring pulses interact due to the fact that their tails decay exponentially fast (see e.g. Chang and Demekhin (2002)) so that the right-hand-side of the corresponding equation in (2.12) for 2D is $-4(\partial/\partial x)H_1(H_2 + H_3)$ with 1 the primary soliton, in 3D we must consider all possible interactions between the three pulses. We can then apply our interaction theory in a straightforward manner to obtain a three-dimensional dynamical system for the locations of the three pulses which can have a chaotic behavior. Such chaotic behavior is frequently observed in interfacial turbulence associated with a falling film (Alekseenko et al., 1994). Similarly, we can extend the theory to n-pulse interaction: for the 3D gKS equation, the right-hand-side is simply $-4(\partial/\partial x)\sum_{i \neq j} H_i H_j$. Nevertheless, as we have demonstrated here, binary interaction can shed light on the pattern formation dynamics of the 3D gKS equation and explain the fundamental feature of the strongly dispersive case, the formation of V-shapes.

The theory can also be extended to any equation. Of particular interest would be the

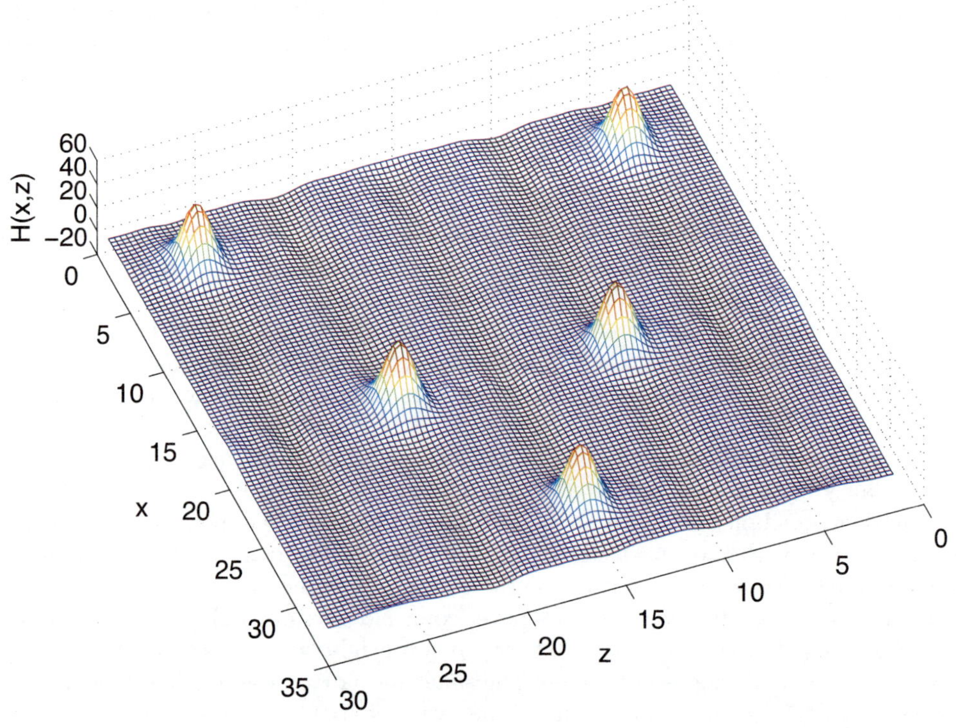

Figure 16. V-shape formation for $\delta = 100$ at $t = 170$ obtained from the 3D gKS equation.

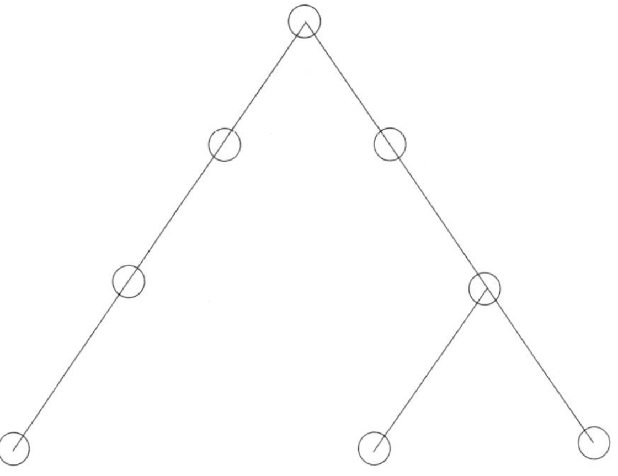

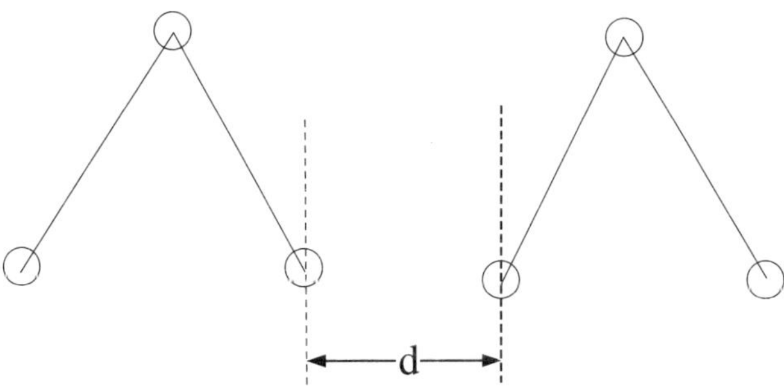

Figure 17. Different possible configurations for a large number of pulses and a sufficiently large domain. The open circles denote 3D solitary humps.

application of the theory to the boundary-layer equations used to describe the dynamics of a falling film for large Reynolds numbers (see e.g. Chang and Demekhin (2002)). Unlike the IBL approximation discussed in §1, the boundary-layer approximation is a higher-level approximation than IBL as it has an additional degree of freedom due to the direction perpendicular to the (x, z) plane. With 3 degrees of spatial freedom we can then expect chaotic behavior for certain regions in the parameter space.

The map of attractors obtained with a generalized n-pulse approach would correspond to an extension of Shilnikov's criterion to 3D as it would give us all possible soliton bound states in 3D by analogy with 2D where the criterion dictates all possible bound states of 2D solitons, e.g. two-hump solitons (Glendinning and Sparrow, 1984). In some sense, the V-shape formation obtained here, offers a partial generalization of Shilnikov's criterion to 3D (as was pointed out above the situation can be more complicated when the number of solitons increases).

Finally, we note that experimental confirmation of the V-shape formation in the strongly dispersive case, i.e. for very small Reynolds numbers (e.g. for very viscous fluids) is still lacking. The V-shaped arrays predicted by the theory have not been observed so far in experimental studies (e.g. Alekseenko et al. (1994, 2005), Tailby and Portalski (1960), Park and Nosoko (2003)), perhaps because the Reynolds numbers in the experiments are typically $O(1)$ or larger. Nevertheless, more careful and detailed experiments need to be carried out to fully decipher the 3D wave regime for the falling film.

Bibliography

S.V. Alekseenko, V.A. Antipin, V.V. Guzanov, S.M. Kharlamov, and D.M. Markovich. Three-dimensional solitary waves on falling liquid film at low reynolds numbers. *Phys. Fluids*, 17:121704, 2005.

S.V. Alekseenko, V.E. Nakoryakov, and B.G. Pokusaev. *Wave Flow of Liquid Films*. Begell House, New York, 1994.

N.J. Balmforth, G.R. Ierley, and E.A. Speigel. Chaotic pulse trains. *SIAM J. Appl. Maths*, 54:1291–1334.

T.B. Benjamin. Wave formation in laminar flow down an inclined plane. *J. Fluid Mech.*, 2:554–574, 1957.

D.J. Benney. Long waves on liquid films. *J. Math. Phys.*, 45:150–155, 1966.

H.-C. Chang. Wave evolution on a falling film. *Annu. Rev. Fluid Mech.*, 26:103–136, 1994.

H.-C. Chang and E.A. Demekhin. Solitary wave formation and dynamics on falling films. *Adv. Appl. Mech.*, 32:1–58, 1996.

H.-C. Chang and E.A. Demekhin. *Complex Wave Dynamics on Thin Films*. Elsevier, New York, 2002.

H.-C. Chang, E.A. Demekhin, and E. Kalaidin. Interaction dynamics of solitary waves on a falling film. *J. Fluid Mech.*, 294:123–154, 1995a.

H.-C. Chang, E.A. Demekhin, and E. Kalaidin. Generation and suppression of radiation by solitary pulses. *SIAM J. Appl. Maths*, 58:1246–1277, 1998.

H.-C. Chang, E.A. Demekhin, and D.I. Kopelevich. Laminarizing effects of dispersion in an active-dissipative nonlinear medium. *Physica D*, 63:299–320, 1993.

H.-C. Chang, E.A. Demekhin, and D.I. Kopelevich. Stability of a solitary pulse against wave packet disturbances in an active medium. *Phys. Rev. Lett.*, 75:1747–1750, 1995b.

K.J. Chu and A.E. Dukler. Statistical characteristics of thin, wavy films. part 2. studies of the substrate and its wave structure. *AIChE J.*, 20:695–706, 1974.

B.I. Cohen, J.A. Krommes, W.M. Tang, and M.N. Rosenbluth. Non-linear saturation of the dissipative trapped-ion mode coupling. *Nucl. Fusion*, 16:971–992, 1976.

E.A. Demekhin, I.A. Demekhin, and V.Ya. Shkadov. Solitons in flowing layer of a viscous fluid. *Izv. Akad. Nauk SSSR, Mekh. Zhidk. Gaza*, 4:9–16, 1983.

E.A. Demekhin and M.A. Kaplan. Construction of exact numerical solutions of the stationary traveling type for viscous thin films. *Izv. Akad. Nauk SSSR, Mekh. Zhidk. Gaza*, 3:23–41, 1989.

E.A. Demekhin, M.A. Kaplan, and V.Ya Shkadov. Mathematical models of the theory of viscous liquid films. *Izv. Akad. Nauk SSSR, Mekh. Zhidk. Gaza*, 6:73–81, 1987.

E.J. Doedel, A.R. Champneys, T.F. Fairgrieve, Y.A. Kuznetsov, B. Sandstede, and X.-J. Wang. *AUTO 97: Continuation and bifurcation software for ordinary differential equations*. Montreal Concordia University (Available via FTP from ftp.cs.concordia.ca in directory pub/doedel/auto), 1997.

C. Elphick, G.R. Ierley, O. Regev, and E.A. Spiegel. Interacting structures with galilean invariance. *Phys. Rev. A*, 44:1110–1122, 1991.

C. Elphick, E. Meron, and E.A. Spiegel. Spatiotemporal complexity in traveling patterns. *Phys. Rev. Lett.*, 61:496–499, 1998.

A.L. Frenkel and K. Indireshkumar. Wavy flows down an inclined plane: Perturbation theory and general evolution equation for the film thickness. *Phys. Rev. E*, 60:4143–4157, 1999.

B. Gjevik. Spatially varying finite-amplitude wave trains on falling liquid films. *Acta Polytech. Scand. Mech. Eng. Ser.*, 61:1–16, 1971.

P. Glendinning and C. Sparrow. Local and global behavior near homoclinic orbits. *J. Stat. Phys.*, 35:645–696, 1984.

D. Gottlieb and S.A. Orszag. *Numerical Analysis of Spectral Methods: Theory and Applications*. SIAM, Philadelphia, 1977.

D.A. Goussis and R.E. Kelly. Surface waves and thermocapillary instabilities in a liquid film flow. *J. Fluid Mech.*, 223:24–45, 1991.

K. Indireshkumar and A.L. Frenkel. Mutually penetrating motion of self-organizing two-dimensional patterns of solitonlike structures. *Phys. Rev. E*, 55:1174–1177, 1997.

A. Joets and R. Ribota. Localized, time-dependent state in the convection of a nematic liquid crystal. *Phys. Rev. Lett.*, 60:2164–2167, 1988.

S.W. Joo, S.H. Davis, and S.G. Bankoff. Long-wave instabilities of heated falling films: two-dimensional theory of uniform layers. *J. Fluid Mech.*, 230:117–146, 1991.

S. Kalliadasis, E.A. Demekhin, C. Ruyer-Quil, and M.G. Velarde. Thermocapillary instability and wave formation on a film falling down a uniformly heated plane. *J. Fluid Mech.*, 492:303–338, 2003a.

S. Kalliadasis, A. Kiyashko, and E.A. Demekhin. Marangoni instability of a thin liquid film heated from below by a local heat source. *J. Fluid Mech.*, 475:377–408, 2003b.

P.L. Kapitza. Wave flow of thin layers of a viscous fluid: I. Free flow. *Zh. Eksp. Teor. Fiz.*, 18:3–28, 1948.

P.L. Kapitza and S.P. Kapitza. Wave flow of thin layers of a viscous fluid: III. Experimental study of undulatory flow conditions. *Zh. Eksp. Teor. Fiz.*, 19:105–120, 1949.

T. Kawahara. Formation of saturated solitons in a nonlinear dispersive system with instability and dissipation. *Phys. Rev. Lett.*, 51:381–383, 1983.

T. Kawahara and S. Toh. Pulse interactions in an unstable dissipative-dispersive nonlinear system. *Phys. Fluids*, 31:2103–2111, 1988.

E.A. Kuznetsov, A.M. Rubechik, and V.E. Zakharov. Soliton stability in plasmas and hydrodynamics. *Phys. Rep.*, 142:103–165, 1986.

S.P. Lin. Finite amplitude side-band stability of a viscous film. *J. Fluid Mech.*, 63:417–429, 1974.

S.P. Lin and M.V.G. Krishna. Stability of liquid film with respect to initially finite three-dimensional disturbances. *Phys. Fluids*, 20:2005–2001, 1977.

C. Nakaya. Waves on a viscous fluid film down a vertical wall. *Phys. Fluids*, 7:1143–1154, 1989.

V.I. Nekorkin and M.G. Velarde. Solitary waves, soliton bound states and chaos in a dissipative korteweg-de vries equation. *Int. J. Bifurcation Chaos*, 4:1135–1146, 1994.

A.A. Nepomnyashchy. Stability of wave regimes in a film flowing down an inclined plane. *Izv. Akad. Nauk SSSR, Mekh. Zhidk. Gaza*, 3:28–34, 1974.

T. Ooshida. Surface equation of falling film flows with moderate Reynolds number and large but finite Weber number. *Phys. Fluids*, 11:3247–3269, 1999.

A. Oron, S.H. Davis, and S.G. Bankoff. Long-scale evolution of thin liquid films. *Rev. Mod. Phys.*, 69:931–980, 1997.

A. Oron and O. Gottlieb. Nonlinear dynamics of temporally excited falling liquid films. *Phys. Fluids*, 14:2622–2636, 2002.

C.D. Park and T. Nosoko. Three-dimensional wave dynamics on a falling film and associated mass transfer. *AIChE J.*, 49:2715–2727, 2003.

J.R.A. Pearson. On convective cells induced by surface tension. *J. Fluid Mech.*, 4: 489–500, 1958.

C.-A. Peng, L.A. Jurman, and M.J. McCready. Formation of solitary waves on gas-sheared liquid layers. *Int. J. Multiphase Flow*, 17:767–782, 1991.

V.I. Petviashvili and O.Yu. Tsvelodub. Horse-shoe shaped solitons on an inclined viscous liquid film. *Dokl. Akad. Nauk SSSR*, 238:1321–1324, 1978.

A. Pumir, P. Manneville, and Y. Pomeau. On solitary waves running down an inclined plane. *J. Fluid Mech.*, 135:27–50, 1983.

B. Ramaswamy, S. Chippada, and S.W. Joo. A full-scale numerical study of interfacial instabilities in thin film flows. *J. Fluid Mech.*, 325:163–194, 1996.

P. Rosenau, A. Oron, and J.M. Hyman. Bounded and unbounded patterns of the Benney equation. *J. Fluid Mech.*, 4:1102–1104, 1992.

G.J. Roskes. Three-dimensional long waves on liquid film. *Phys. Fluids*, 13:1440–1445, 1970.

C. Ruyer-Quil and P. Manneville. Improved modeling of flows down inclined planes. *Eur. Phys. J. B*, 15:357–369, 2000.

C. Ruyer-Quil and P. Manneville. Further accuracy and convergence results of the modeling of flows down inclined planes by weighted residual approximations. *Phys. Fluids*, 14:170–183, 2002.

C. Ruyer-Quil, B. Scheid, S. Kalliadasis, M.G. Velarde, and R.Kh Zeytounian. Thermocapillary long waves in a liquid film flow. Part 1. Low-dimensional formulation. *J. Fluid Mech.*, 538:199–222, 2005.

T.R. Salamon, R.C. Armstrong, and R.A. Brown. Traveling waves on vertical films: Numerical analysis using the finite element method. *Phys. Fluids*, 6:2202–2220, 1993.

S. Saprykin, E.A. Demekhin, and S. Kalliadasis. Two-dimensional wave dynamics in thin films. Part I. Stationary solitary pulses. *Phys. Fluids*, 17:117105, 2005.

B. Scheid. *Evolution and Stability of Falling Liquid Films with Thermocapillary Effects. PhD Thesis*. Université Libre de Bruxelles, 2004.

B. Scheid, A. Oron, P. Colinet, U. Thiele, and J.C. Legros. Nonlinear evolution of nonuniformly heated falling liquid films. *Phys. Fluids*, 14:4130–4151, 2002.

B. Scheid, C. Ruyer-Quil, S. Kalliadasis, M.G. Velarde, and R.Kh Zeytounian. Thermocapillary long waves in a liquid film flow. Part 2. Linear stability and nonlinear waves. *J. Fluid Mech.*, 538:223–244, 2005a.

B. Scheid, C. Ruyer-Quil, U. Thiele, O.A. Kabov, J.C. Legros, and P. Colinet. Validity domain of the Benney equation including Marangoni effect for closed and open flows. *J. Fluid Mech.*, 527:303–335, 2005b.

D.R. Scott and D.J. Stevenson. Magma ascent by porous flow. *J. Geophys. Res.*, 91: 9283–9296, 1986.

L.E. Scriven and C. Sternling. On cellular convection driven by surface tension gradients. *J. Fluid Mech.*, 19:321–340, 1964.

V.Ya Shkadov. Wave models in the flow of a thin layer of a viscous liquid under the action of gravity. *Izv. Akad. Nauk SSSR, Mekh. Zhidk. Gaza*, 1:43–50, 1967.

V.Ya Shkadov. Theory of wave flow of a thin layer of a viscous liquid. *Izv. Akad. Nauk SSSR, Mekh. Zhidk. Gaza*, 2:20–25, 1968.

V.Ya Shkadov. *Some Methods and Problems of the Theory of Hydrodynamic Stability.* Institute of Mechanics, Moscow, 1973.

V.Ya Shkadov. Solitary waves in a layer of viscous liquid. *Izv. Akad. Nauk SSSR, Mekh. Zhidk. Gaza*, 12:63–68, 1977.

S.R. Tailby and S. Portalski. The hydrodynamics of liquid films flowing on vertical surface. *Trans. Instn Chem. Engrs*, 38:324–330, 1960.

S. Toh, H. Iwasaki, and T. Kawahara. Two-dimensionally localized pulses of a nonlinear equation with dissipation and dispersion. *Phys. Rev. A*, 40:5472–5475, 1989.

J. Topper and T. Kawahara. Approximate equation for long nonlinear waves on viscous fluid. *J. Phys. Soc. Jpn*, 44:663–666, 1978.

P.M.J. Trevelyan and S. Kalliadasis. Dynamics of a reactive falling film at large Péclet numbers. I. Long-wave approximation. *Phys. Fluids*, 16:3191–3208, 2004a.

P.M.J. Trevelyan and S. Kalliadasis. Dynamics of a reactive falling film at large Péclet numbers. II. Nonlinear waves far from criticality: Integral-boundary-layer approximation. *Phys. Fluids*, 16:3209–3226, 2004b.

P.M.J. Trevelyan and S. Kalliadasis. Wave dynamics on a thin-liquid film falling down a heated wall. *J. Eng. Math.*, 50:177–208, 2004c.

P.M.J. Trevelyan, B. Scheid, C. Ruyer-Quil, and S. Kalliadasis. Heated falling films. *J. Fluid Mech.*, submitted, 2006.

R. Usha and B. Uma. Modeling of stationary waves on a thin viscous film down an inclined plane at high Reynolds numbers and moderate Weber numbers using energy integral method. *Phys. Fluids*, 16:2679–2696, 2004.

C.-S. Yih. Stability of liquid flow down an inclined plane. *Phys. Fluids*, 6:321–334, 1963.

V.E. Zakharov and E.A. Kuznetsov. On three-dimensional solitons. *Sov. Phys. JETP*, 39:285–286, 1974.

Miscible Fingering in Electrokinetic Flow: Symmetries and Zero Modes

Yuxing Ben[*†], Evgeny A. Demekhin[‡] and Hsueh-Chia Chang[‡+]

[*] Department of Mathematics, Massachusetts Institute of Technology, Cambridge, MA, USA

[†] Institute for Soldier Technologies, Massachusetts Institute of Technology, Cambridge, MA, USA

[‡] Department of Chemical and Biomolecular Engineering, University of Notre Dame, Notre Dame, IN, USA

[+] Center for Microfluidics and Medical Diagnostics, University of Notre Dame, Notre Dame, IN, USA

Abstract Analysis is carried out to examine the miscible fingering phenomenon when a fluid with a lower electrolyte concentration advances into one with a higher concentration. Unlike earlier miscible fingering theories, the linear stability analysis is carried out in the self-similar coordinates of the diffusing front. The dominant destabilizing mode is shown to be a localized concentration disturbance within the diffusive layer. However, transverse diffusion is shown to eventually stabilize the instability at time $t = t_c$ that scales as $b^{\frac{8}{3}}$, where b is the breadth of the channel, and lengthen the dominant transverse wavelength as $t^{\frac{3}{8}}$ in the interim. With these scaling laws, we arrive at the conclusion that transient fingering is insignificant in sub-millimeter micro-devices which use electrokinetic flow to transport biological and environmental electrolyte samples.

1 Introduction

Extended-domain spatial-temporal chaos is, by definition, a complex dynamical system with many degrees of freedom and a continuum of length and time scales. A simple description based on low-dimensional dynamics is only possible if there exist robust and localized coherent structures that persist during the dynamics. The many degrees of freedom would then correspond to the number of such structures and the continuum of length and time scales to the widely varying separations and interaction time scales between them. Despite the large number of coherent structures, their interaction dynamics is often described by only a few dominant eigenmodes of the localized structure. These dominant modes result from certain physical symmetries of the structures and can often be identified a priori. As a result, the most important statistics of complex spatial-temporal dynamics can indeed be extrapolated from a low dimensional description of the interaction dynamics between two neighboring structures via a mean-field theory. We have successfully described complex wave dynamics (interfacial turbulence) on a falling film with this approach. However, we shall avoid repeating the material in our recent book on the subject (Chang and Demekhin, 2002). While the content of the book was reviewed in the lectures, we endeavor here to tackle a different extended-domain dynamics using the same approach. We shall again identify certain key symmetries in the problem and use them to determine the dominant zero modes. This new phenomenon is miscible fingering in electrokinetic flow and this chapter serves as an example to our approach to extended domain dynamics. It also links our past research on film waves, which was the focus of the lectures, to our current research interests summarized on our website (www.nd.edu/ changlab/).

Electrokinetic flow has become a promising means of transporting electrolyte samples in micro-devices. Unlike pressure-driven flow, it offers a flat velocity profile in the small flow channels of these devices. As a result, Taylor dispersion, a macroscopic axial mixing mechanism that results from a coupling between transverse diffusion and transverse velocity gradient, is minimized (Taylor, 1954; Aris, 1956). In so far as this dispersion mechanism is the dominant mixing phenomenon in micro-channels, its reduction allows smaller sample spacing and higher throughput – an important consideration for future multi-sample devices (Culbertson et al., 1998; Ermakov et al., 1998; Freemantle, 1999). Electrokinetic flow is also easier to manipulate and control in a multi-channel device, as electrodes can be easily inserted into the micro-circuitry technology. A smart micro-laboratory-on-a-chip, which can direct the sample to different downstream stations depending on the analytic outcome at a particular station, has been envisioned by new start-up microfluidic companies like Caliper Inc.

There are, however, many obstacles that must be overcome before electrokinetic flow becomes a reliable fluid and sample transport mechanism for future micro-devices. Protein can denature under the high electric fields of electrokinetic flow. Non-electrolyte samples and gaseous reactants require a new transport strategy if electrokinetic driving forces are used (Takhistov et al., 2002).

In this article, we address another potential problem in electrokinetic flow – miscible fingering. The electrolyte sample will significantly increase the local electrolyte concentration relative to the carrier buffer electrolyte concentration. As the electroosmotic slip velocity decreases with increasing concentration C as $C^{-\frac{1}{2}}$, the carrier suffers a smaller hydrodynamic resistance than the sample. A pressure-driven back flow will hence be established within the carrier so that the sample and carrier can move at the same velocity (Herr et al., 2000) to ensure mass conservation. However, such flows are expected to be unstable at the back end of the sample. Any local advance of the high-mobility carrier would be accelerated due to its favorable mobility to produce a carrier finger into the sample. Due to mass conservation, a similar protrusion of the sample can then result in the front or a back sample finger can appear at the back – both leading to significant increase in sample length and can potentially induce sample mixing like Taylor dispersion. Unlike their electrokinetic counterpart, viscous and gravitational miscible fingering have been studied for many decades. Such fingering is irreversible and a characteristic transverse wavelength is observed at onset. A schematic illustration of miscible fingering is shown in Figure 1. However, unlike immiscible fingering which undergoes extensive tip splitting to form smaller and smaller fingers, the dominant transverse wavelength increases in time in miscible fingering. Explanation for both the selected wavelength at onset and its subsequent coarsening is still lacking. An effec-

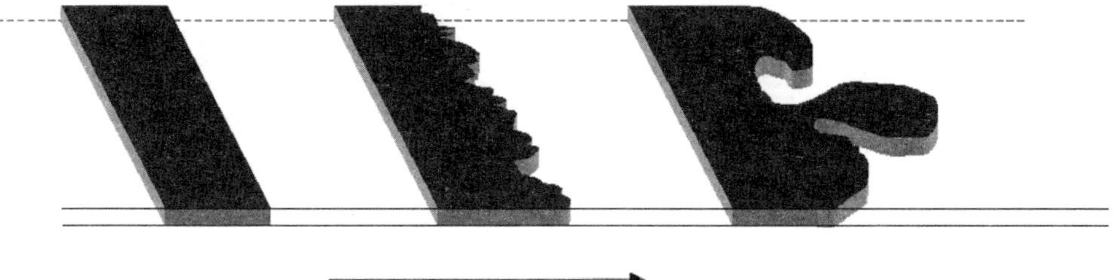

Figure 1. A schematic illustration of miscible fingering. Fingering coarsens due to diffusion.

tive surface tension for miscible fluids is sometimes inserted to stabilize short waves and select long ones (Hu and Joseph, 1992). However, such surface tension is negligible and often undefinable for miscible fluids. Transverse diffusion has also been suggested as a mechanism for wavelength selection. Indeed, analyses which include transverse diffusion yield a longwave instability with a selected wavelength (Yortsos and Zeybek, 1988; Tan and Homsy, 1986). However, these earlier theories are unable to quantitatively capture the subsequent coarsening. They often invoke two major assumptions – a quasi-steady approximation assuming downstream diffusion of the front is slow compared to the finger growth and a normal mode approximation where the disturbances are cast in the form of Fourier modes in both the downstream and transverse directions. Both approximations are made to produce a classical Orr-Sommerfeld type equations with pseudo-constant coefficients and with normal mode solutions. Both approximations are incorrect. Since the thickness of the diffusion layer grows as $\sqrt{Dt}$, the concentration gradient that drives the instability decays as $t^{-\frac{1}{2}}$. As a result, the instability is most dominant at small t when the gradient is fast varying. In contrast, as with all longwave instability, the growth rate is rather slow – just the opposite of the invoked assumption. Moreover, one expects the disturbances to be localized within the diffusive layer, in direct contradiction with the global nature of the assumed Fourier mode in the streamwise direction.

We shall develop a new miscible fingering theory for electrokinetic flow here, although our approach can be applied to other classical miscible fingering phenomena. In Figure 2, we show photographic images of our preliminary experiments for electrokinetic fingering. A potassium permaganate ($KMnO_4$) solution of concentration 10^{-2}mol/l is added over a patch of NaCl buffer solution of concentration 10^{-2} mol/l in a horizontal slot of 1 mm in gap width, 10 cm wide and 10 cm long. Since the solution is denser than the buffer, the slot is first tilted in the direction of the $KMnO_4$ solution until a clear linear boundary is discernible between the solution and the buffer. The slot is then returned to its horizontal position and an electric field of 10 V/cm is applied to drive the fluid from the buffer to the solution – the unstable configuration of a high-mobility fluid displacing a low-mobility one. This electric field is comparable to that used in micro-devices. Since the electrokinetic velocity is independent of the gap width, provided it is larger than the sub-micron double layer thickness, the front displacement velocity is also compatible to that in micro-devices. The dye front is displaced from left to right

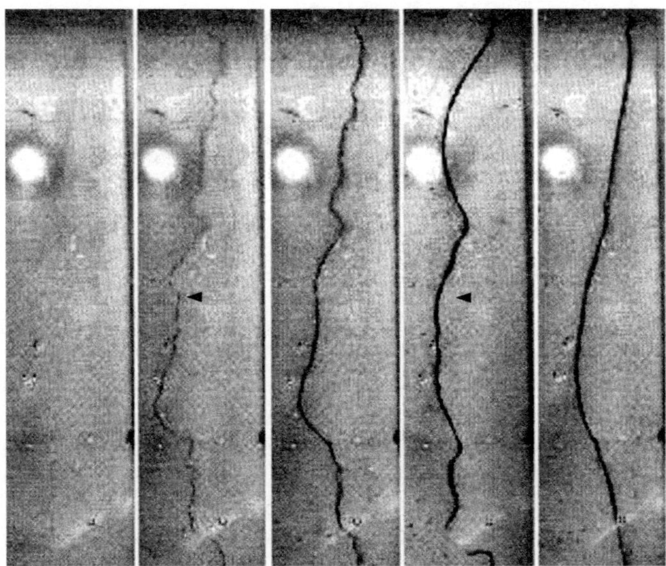

Figure 2. Photographic images of an electrokinetic phenomenon with $KMnO_4$ concentration of 10^{-2} mol/m, NaCl buffer concentration of 10^{-2} mol/l, slot gap of 1mm, slot width and length are 10cm and the field is 10V/cm. The frames are taken at an interval of 2.5 minutes. The front advances from the left to the right in each frame. The sharper interface in the later frames is due to a precipitation reaction of the potassium permaganate. Fingering and subsequent coarsening are evident at the front. This second set disappears in the later frames.

in each frame. The $KMnO_4$ dye is weak and the color contrast across the front is not sharp in the first few frames of Figure 2. However, an OH^- front also propagates from the cathode on the right to the anode on the left. When this front meets the $KMnO_4$ front, an oxidation reaction results and produces salt precipitates. Consequently, the dye fronts in later frames are actually sharper despite downstream diffusion – the intensity is enhanced by the salt precipitates. Despite the light initial contrast, fingering instability is evident from the outset in the initial frames. As is evident, short saw-tooth fluctuations appear immediately at the boundary and coarsen in time. However, unlike other miscible fingering, this instability is transient and disappears after some interval in time ($\sim$20 minutes). This is extremely desirable for micro-device applications as no prolonged fingering occurs. However, both the transient nature and the coarsening law with respect to time must be verified theoretically. Twenty minutes is much larger than the residence time of most micro-devices and the transient instability can still corrupt the efficiency and quality of the device.

2 Governing Equations

We shall first derive the depth-averaged mass transport and momentum equations for electrokinetic flow in a slot. We assume that the buffer solution (NaCl) has a uniform concentration. Another electrolyte (KMnO$_4$) is added at one end of the channel. If K$^+$ is denoted by the subscript $+$ and MnO$_4^-$ by the subscript $-$, the mass balances of individual ions can be written in the form

$$\frac{\partial c_+}{\partial t} + \mathbf{u} \cdot \nabla c_+ = z_+ v_+ F \nabla \cdot (c_+ \nabla \phi) + D_+ \nabla^2 c_+ \tag{2.1}$$

$$\frac{\partial c_-}{\partial t} + \mathbf{u} \cdot \nabla c_- = z_- v_- F \nabla \cdot (c_- \nabla \phi) + D_+ \nabla^2 c_- . \tag{2.2}$$

Here, c_+ and c_- are the concentrations of K$^+$ and MnO$_4^-$ respectively, v the mobility, which is related to the diffusivity D by the relation $D = RTv$, z the charge number of the species, $\mathbf{u}$ the velocity, ϕ the electric potential, and F the Faraday constant equal to the charge of 1 mol of single-valenced ions, $F = 9.65 \times 10^4 \text{C mol}^{-1}$.

The condition of electroneutrality then requires

$$z_+ c_+ + z_- c_- = 0. \tag{2.3}$$

For our single-valenced K$^+$MnO$_4^-$, $z_\pm = \pm 1$ and we have $c_+ = c_-$. Hence, we can replace c_+ and c_- by c. Subtracting the equation for the negative ions from that for the positive ions gives

$$(z_+ v_+ - z_- v_-) F \nabla \cdot (c \nabla \phi) + (D_+ - D_-) \nabla^2 c = 0. \tag{2.4}$$

This relation may be used to eliminate the potential from either the positive ion or negative ion convection-diffusive equation to give

$$\frac{\partial c}{\partial t} + \mathbf{u} \cdot \nabla c = D \nabla^2 c. \tag{2.5}$$

Here, D is an effective diffusion coefficient defined by $D = \frac{2 D_+ D_-}{D_+ + D_-}$.

Because of the motion of the charged species, there will be a current. Specially, the current density $\mathbf{i}$ is given by

$$\mathbf{i} = -F^2 \nabla \phi \sum z_i^2 v_i c_i - F \sum z_i D_i \nabla c_i + F \mathbf{u} \sum z_i c_i \tag{2.6}$$

or

$$\mathbf{i} = -\sigma \nabla \phi - F \sum z_i D_i \nabla c_i + F \mathbf{u} \sum z_i c_i \tag{2.7}$$

where $\sigma = F^2 \sum z_i^2 v_i c_i$ is the scalar electrical conductivity of the solution. Equation (2.3) states that there is no local accumulation of charge which corresponds to $\nabla \cdot \mathbf{i} = 0$. In addition, the convective contribution $F \mathbf{u} \sum z_i c_i$ to the current density in equation (2.7) vanishes. In the following analysis, we assume the diffusivities of counter and co-ions to be similar in magnitude and their net diffusion contribution to the current is assumed to vanish due to mutual cancelation. Hence, we can neglect the second term of the equation above. We get

$$\mathbf{i} = -\sigma \nabla \phi \tag{2.8}$$

or with current balance

$$\nabla \cdot (\sigma \nabla \phi) = 0. \tag{2.9}$$

The conductivity for the present case can be written as

$$\sigma = F^2 (v_{\mathrm{Na}^+} c_0 + v_{\mathrm{K}^+} c + v_{\mathrm{Cl}^-} c_0 + v_{\mathrm{MnO}_4^-} c) \tag{2.10}$$

where c_0 is the concentration of the buffer solution (NaCl) and c is the concentration of our dye (KMnO$_4$). The buffer solution c_0 is constant as assumed, which is roughly $10^{-2} mole/$l. Then the current balance equation (9) becomes

$$(c + \beta c_0)\nabla^2 \phi + \nabla c \cdot \nabla \phi = 0 \tag{2.11}$$

where $\beta = \frac{D_{\mathrm{Na}^+} + D_{\mathrm{Cl}^-}}{D_{\mathrm{K}}^+ + D_{\mathrm{MnO}_4^-}} = 0.938$.

With the electric body force per unit volume given by $\mathbf{f}_E = \rho_E \mathbf{E}$, the momentum equation may be written as

$$\mu \nabla^2 \mathbf{u} = \nabla p + \rho_E \mathbf{E}. \tag{2.12}$$

Here, μ is the viscosity of the fluid, ρ_E is the electric charge density (Cm^{-3}), and $\mathbf{E} = -\nabla \phi$ is the electric field.

We designate x as the coordinate directed along the axis towards the cathode, u the velocity component in that direction and v the velocity component in the transverse direction. For the basic state front without transverse disturbances, we have locally linear pressure and electric potential profiles for a straight front in a slot. If, in addition, the double layer thickness λ_D is small compared with the channel width, we can carry out a depth averaging of the velocity in the transverse x and y direction to yield

$$u = -\frac{d^2}{12\mu}\frac{\partial p}{\partial x} - \frac{\epsilon}{\mu}\zeta(c)\frac{\partial \phi}{\partial x} \tag{2.13}$$

$$v = -\frac{d^2}{12\mu}\frac{\partial p}{\partial y} - \frac{\epsilon}{\mu}\zeta(c)\frac{\partial \phi}{\partial y} \tag{2.14}$$

where d is the depth of the channel, $\zeta = \frac{k_3}{\sqrt{c+c_0}}$, which captures how the zeta potential ζ and the electroosmotic velocity scales as the $-\frac{1}{2}$ power of the total concentration (the KMnO$_4$ concentration c plus the buffer concentration c_0), and $k_3 = q(\frac{\epsilon\epsilon_0 kT}{2e^2})^{\frac{1}{2}}$ from a thermodynamic analysis of the double layer (Probstein, 1994). The parameter k is the Boltzmann constant, q the total charge in the double layer, ϵ the permittivity of a vacuum, ϵ_0 the relative permittivity, T the temperature, and e the elementary charge.

The continuity equation

$$\frac{\partial u}{\partial x} + \frac{\partial v}{\partial y} = 0 \tag{2.15}$$

and the boundary conditions

$$\phi = 0, \ p = 0, \ c = c_2, \ \text{at } x = 0 \tag{2.16}$$

$$\phi = \phi_0, \ p = 0, \ c = c_1, \ \text{at } x = w \tag{2.17}$$

where w is the length of the channel, then complete the mass and momentum transport equations.

We non-dimensionalize these equations as follows:

$$\tilde{c} = \frac{c}{c_0}, \quad \tilde{\phi} = \frac{\phi}{\phi_0}, \quad \tilde{u} = \frac{u}{U_0} \tag{2.18}$$

$$\tilde{x} = \frac{x}{L}, \quad \tilde{y} = \frac{y}{L}, \quad \tilde{t} = \frac{t}{T}, \quad \tilde{p} = \frac{p}{\mu \frac{U_0}{L}}. \tag{2.19}$$

Here, $U_0 = \frac{\epsilon k_3}{\mu \sqrt{c_0}} \frac{\phi_0}{w}$ is the electrokinetic velocity for an electrolyte of concentration c_0 due to a potential drop ϕ_0 over a length w. Because fingering occurs at the diffusive layer, we choose the characteristic length as $L = \frac{D}{U_0}$ and the characteristic time as $T = \frac{D}{U_0^2}$. For convenience, we drop the tildes from the dimensionless equations and obtain:

$$\frac{\partial c}{\partial t} + u\frac{\partial c}{\partial x} + v\frac{\partial c}{\partial y} = \frac{\partial^2 c}{\partial x^2} + \frac{\partial^2 c}{\partial y^2} \tag{2.20}$$

$$u = -\gamma\frac{\partial p}{\partial x} - \frac{Pe}{\sqrt{1+c}}\frac{\partial \phi}{\partial x} \tag{2.21}$$

$$v = -\gamma\frac{\partial p}{\partial y} - \frac{Pe}{\sqrt{1+c}}\frac{\partial \phi}{\partial y} \tag{2.22}$$

$$\frac{\partial u}{\partial x} + \frac{\partial v}{\partial y} = 0 \tag{2.23}$$

$$(c + \beta)\nabla^2\phi + \nabla c \cdot \nabla\phi = 0 \tag{2.24}$$

Here, $Pe = \frac{U_0 w}{D} = \frac{\epsilon k_3 \phi_0}{\mu D \sqrt{c_0}}$, and $\gamma = \frac{1}{12}\frac{d^2}{w^2}Pe^2$. Since we are considering the diffusing front, the solutions we obtain from the equations above are actually the inner solutions. Their boundary conditions at the infinities need to match the outer solutions.

3 Basic State

Suppose that the position of the diffusing front is at $l(t)$. The front values ϕ_f and p_m are the electric potential and pressure, respectively, at $l(t)$. The outer solution for the concentration is

$$c = c_1 \text{ at } x \geq l(t); \quad c = c_2 \text{ at } x < l(t). \tag{3.1}$$

From equation (2.9), the electric current i_x must be constant with respect to x, hence

$$i_x = -\sigma_2 \frac{\phi_f - \phi_0}{l(t)} = -\sigma_1 \frac{0 - \phi_f}{w - l(t)}. \tag{3.2}$$

We then obtain

$$\phi_f = \frac{\phi_0}{1 + \frac{\sigma_1}{\sigma_2}\frac{l}{w-l}} \tag{3.3}$$

where $\sigma_1 = c_1 + \beta$, $\sigma_2 = c_2 + \beta$. Substituting ϕ_f into $\mathbf{i}$ results in the steady current

$$i_x = I = \frac{(-\phi_0)\sigma_1\sigma_2}{(w-l)\sigma_2 + \sigma_1 l}. \tag{3.4}$$

Although the front position $l(t)$ changes in time, it does so slowly and the current I will be assumed constant.

Near $x = 0^-$, or in other words, on the left side of the diffusion layer the electrokinetic velocity is:

$$U = -\frac{Pe}{\sqrt{1+c_2}}\frac{\phi_f - \phi_0}{l} - \gamma\frac{p_m}{l}. \tag{3.5}$$

Likewise, on the right side of the diffusion layer, the electrokinetic velocity is :

$$U = -\frac{Pe}{\sqrt{1+c_1}}\frac{\phi_f}{w-l} + \gamma\frac{p_m}{w-l}. \tag{3.6}$$

The continuity equation implies that the flow rate is constant along a streamline. We hence have that the velocity is equal to that outside the diffusion layer since it is constant on both sides of it. Subtracting equation (3.5) from equation (3.6) produces the pressure at the front

$$p_m = \frac{\frac{Pe}{\gamma}\left(\frac{\phi_0-\phi_f}{\sqrt{1+c_2}} - \frac{\phi_f}{\sqrt{1+c_1}}\frac{l}{w-l}\right)}{1 + \frac{l}{w-l}}. \tag{3.7}$$

From the analysis above, we know that the flat front velocity U is constant. Let us then change the coordinate system by $z = x - Ut$ such that the diffusive layer is stationary in z. In this reference frame, the transport equation (2.20) becomes a diffusion equation in x with the usual self-similar solution

$$C_0 = \frac{c_1 + c_2}{2} + \frac{c_1 - c_2}{2}\mathrm{erf}\left(\frac{z}{\sqrt{4t}}\right) \tag{3.8}$$

with $\mathrm{erf}(\eta) = \frac{2}{\sqrt{\pi}}\int_0^\eta e^{-x^2}dx$. The current balance (2.24) implies that the current is constant within the inner diffusive layer and must be identical to that in the outer solution (3.4),$(c+\beta)\frac{\partial\phi}{\partial x} = I$. Hence, the inner potential gradient for the basic state is

$$\frac{d\Phi}{dz} = D\Phi_0 = \frac{I}{C_0 + \beta}. \tag{3.9}$$

4 Self-Similar Linear Stability Analysis

We assume that the disturbed solutions have the following normal mode form

$$c = C_0 + \hat{c}(z,t)e^{i\alpha y}, u = \hat{u}(z,t)e^{i\alpha y}, v = \hat{v}(z,t)e^{i\alpha y}, \Phi = \Phi_0 + \hat{\phi}(z,t)e^{i\alpha y}, \tag{4.1}$$

corresponding to a Fourier transform in the y direction. The perturbed equations (2.20)-(2.24) are then

$$\frac{\partial\hat{c}}{\partial t} + \hat{u}\cdot DC_0 = \frac{\partial^2\hat{c}}{\partial z^2} - \alpha^2\hat{c} \tag{4.2}$$

$$\hat{u} = -\gamma \frac{\partial \hat{p}}{\partial z} - \frac{Pe}{\sqrt{1+C_0}} \frac{\partial \hat{\phi}}{\partial z} + \frac{Pe \cdot D\Phi_0}{2(1+C_0)^{\frac{3}{2}}} \hat{c} \tag{4.3}$$

$$\hat{v} = -i\alpha\gamma\hat{p} - i\alpha \frac{Pe}{\sqrt{1+C_0}} \hat{\phi} \tag{4.4}$$

$$\frac{\partial \hat{u}}{\partial z} + i\alpha\hat{v} = 0 \tag{4.5}$$

$$\frac{\partial^2 \hat{\phi}}{\partial z^2} - \alpha^2 \hat{\phi} + \frac{DC_0}{(C_0+\beta)} \frac{\partial \hat{\phi}}{\partial z} + \frac{D\Phi_0}{(C_0+\beta)} \frac{\partial \hat{c}}{\partial z} - \frac{D\Phi_0 DC_0}{(C_0+\beta)^2} \hat{c} = 0 \tag{4.6}$$

where $DC_0 = \frac{dC_0}{dz}$. Equations (3.8),(3.9) and (4.3) can also be combined to eliminate the disturbance pressure $\hat{p}$ and the transverse velocity $\hat{v}$

$$\frac{\partial^2 \hat{u}}{\partial z^2} - \alpha^2 \hat{u} + \frac{\alpha^2 Pe}{(1+C_0)^{\frac{3}{2}}} \left(-\frac{1}{2} DC_0 \cdot \hat{\phi} + \frac{1}{2} D\Phi_0 \cdot \hat{c} \right) = 0. \tag{4.7}$$

Equations (4.2),(4.6) and (4.7) are the final perturbed equations needed to be solved.

One notices that the basic state for the concentration is self-similar which suggests that the disturbed equations also have some similarity symmetry. Therefore, we transform the problem to the self-similar coordinates where

$$\tau = t, \quad \xi = \frac{x}{\sqrt{4t}}. \tag{4.8}$$

The equations then become:

$$\frac{\partial \hat{c}}{\partial \tau} - \frac{1}{2} \frac{\xi}{\tau} \frac{\partial \hat{c}}{\partial \xi} - \frac{1}{4\tau} \frac{\partial^2 \hat{c}}{\partial \xi^2} + \alpha^2 \hat{c} + a_1 \hat{u} = 0 \tag{4.9}$$

$$\frac{\partial^2 \hat{\phi}}{\partial \xi^2} \frac{1}{4\tau} - \alpha^2 \hat{\phi} + a_2 \frac{1}{\sqrt{4\tau}} \frac{\partial \hat{\phi}}{\partial \xi} - a_3 \frac{\partial \hat{c}}{\partial \xi} \frac{1}{\sqrt{4\tau}} + a_4 \hat{c} = 0 \tag{4.10}$$

$$\frac{\partial^2 \hat{u}}{\partial \xi^2} \frac{1}{4\tau} - \alpha^2 \hat{u} - Pe a_5 \hat{\phi} + Pe a_6 \hat{c} = 0. \tag{4.11}$$

Here,

$$a = \frac{c_1 - c_2}{2\sqrt{\pi}}, \quad a_1 = \frac{1}{\tau^{\frac{1}{2}}} a e^{-\xi^2}, \quad a_2 = \frac{1}{\tau^{\frac{1}{2}}} a e^{-\xi^2} \frac{1}{(C_0+\beta)} \tag{4.12}$$

$$a_3 = -\frac{D\Phi_0}{(C_0+\beta)}, \quad a_4 = -\frac{D\Phi_0}{(C_0+\beta)^2} \frac{1}{\tau^{\frac{1}{2}}} a e^{-\xi^2} \tag{4.13}$$

$$a_5 = \frac{1}{2} \frac{\alpha^2}{(1+C_0)^{\frac{3}{2}}} \frac{1}{\tau^{\frac{1}{2}}} a e^{-\xi^2}, \quad a_6 = \frac{1}{2} \frac{\alpha^2}{(1+C_0)^{\frac{3}{2}}} D\Phi_0. \tag{4.14}$$

The appropriate boundary conditions for equations (4.9)-(4.11) are: $\hat{\phi} = \hat{c} = \hat{u} = 0$, at $\xi = \pm\infty$.

In order to get the proper eigenfunctions in the following analysis, another transformation with $\hat{\theta} = \sqrt{\tau}\hat{c}$ yields

$$\frac{\partial^2 \hat{\theta}}{\partial \xi^2} + 2\xi \frac{\partial \hat{\theta}}{\partial \xi} + 2\hat{\theta} = 4\tau \frac{\partial \hat{\theta}}{\partial \tau} + 4\alpha^2 \tau \hat{\theta} + 4ae^{-\xi^2}\tau \hat{u} \qquad (4.15)$$

$$\frac{\partial^2 \hat{\phi}}{\partial \xi^2} - 4\tau\alpha^2 \hat{\phi} + 2ae^{-\xi^2}\frac{1}{(C_0 + \beta)}\frac{\partial \hat{\phi}}{\partial \xi} + \frac{2D\Phi_0}{(C_0 + \beta)}\frac{\partial \hat{\theta}}{\partial \xi} - 4ae^{-\xi^2}\frac{D\Phi_0}{(C_0 + \beta)^2}\hat{\theta} = 0 \qquad (4.16)$$

$$\frac{\partial^2 \hat{u}}{\partial \xi^2} - 4\tau\alpha^2 \hat{u} - 2Pe\alpha^2\sqrt{\tau}G(\xi)\hat{\phi} - 4\sqrt{\tau}Pe\alpha^2 H(\xi)\hat{\theta} = 0 \qquad (4.17)$$

where $G(\xi) = \frac{ae^{-\xi^2}}{(1+C_0)^{\frac{3}{2}}}$ and $H(\xi) = \frac{-I}{2(1+C_0)^{\frac{3}{2}}(C_0+\beta)}$.

There are several reasons for casting the perturbation equations as (4.15) to (4.17). We seek localized solutions that vanish at $\xi = \pm\infty$ and hence the resulting eigenfunctions associated with the dominant operator of the equations must be localized. This is ensured by the diffusion operator on the left-side of (4.15) in the limit of $\tau \to 0$. In this limit, the diffusing front has a fast-decaying potential gradient $D\Phi_0$ and hence cannot be captured by a quasi-steady approximation. The localized eigenfunction is also far from the assumed normal mode in previous theories. Moreover, we require a longwave instability with a zero growth rate at $\alpha = 0$. We would then carry out a longwave expansion in α. This requires the same dominant operator to be singular with a valid null adjoint eigenfunction to allow the solvability condition to be invoked in the expansion. This is again satisfied by the dominant diffusion operator on the left of equation (4.15).

Equation (4.15) can be written as

$$\tau\hat{\theta}_\tau = L\hat{\theta} + f(\hat{\theta}, \alpha, \hat{u}) \qquad (4.18)$$

where the dominant diffusion operator becomes

$$L\hat{\theta} = \hat{\theta}_{\xi\xi} + 2\xi\hat{\theta}_\xi + 2\hat{\theta}. \qquad (4.19)$$

Now the linear operator L has a discrete spectrum of

$$\lambda = -\frac{n}{2}, \ n = 0, 1, 2, \cdots \qquad (4.20)$$

with localized eigenfunctions of the form $e^{\lambda_n \tau - \xi^2} H_n(\xi)$, where $H_n(\xi)$ are Hermite polynomials. Hence, this operator is singular. It can be readily shown that the perturbation $f(\hat{\theta}, \alpha, \hat{u})$ vanishes in the limit of $\alpha \to 0$ as $\hat{u}$ also vanishes in the same limit. Hence, the instability must be a longwave one that perturbs the null eigenvalue $\lambda_0(\alpha)$ from zero for finite α.

That the null discrete mode with eigenvalue λ_0 is the selected one requires more in-depth analysis. There are essential eigenfunctions to (4.19) that do not decay at $\xi = \pm\infty$. However, we find numerically that all such modes have negative eigenvalues and hence cannot compete with the discrete null mode corresponding to λ_0. Hence, after some transient, the dominant concentration mode is $\hat{\theta} = A(\tau; \alpha)e^{-\xi^2}$. We then focus on the large-time behavior of this mode which is the only surviving one after the transient. Note there is actually a continuum of this discrete mode parameterized by the transverse wavenumber α. We seek the most dominant one over α at large τ.

Equations (4.15), (4.16) and (4.17) are coupled with each other. Fortunately, we find that the third term in (4.17) does not contribute much compared with the fourth term. We first define $\xi = \frac{y}{2\alpha\sqrt{\tau}}$ such that (4.16) yields

$$\frac{\partial^2 \hat{\phi}}{\partial y^2} - \hat{\phi} + \frac{2ae^{-\xi^2}}{(C_0 + \beta) \cdot 2\alpha\sqrt{\tau}}\frac{\partial \hat{\phi}}{\partial y} + \frac{2D\phi_0}{(C_0+\beta)c\dot{\alpha}\sqrt{\tau}}\frac{\partial \hat{\theta}}{\partial \xi} - 4ae^{-\xi^2}\frac{D\phi_0}{(C_0+\beta)^2(2\alpha\sqrt{2})^2}\hat{\theta} = 0.$$

$$(4.21)$$

From this equation, it is clear that the disturbance potential $\hat{\phi}$ can be expanded as $\hat{\phi} = \hat{\phi}_0 + \frac{1}{\alpha\sqrt{\tau}}\hat{\phi}_1$ for large τ. The leading-order equation is just the Fourier-transformed Laplace equation

$$\frac{\partial^2 \hat{\phi}_0}{\partial y^2} - \hat{\phi}_0 = 0. \tag{4.22}$$

With vanishing boundary conditions at infinity, the leading-order solution hence vanishes exactly. This means that the conductivity variance cannot produce a change in $\hat{\phi}$ at large time. For the next order

$$\frac{\partial^2 \hat{\phi}_1}{\partial y^2} - \hat{\phi}_1 = -\frac{D\phi_0}{(C_0+\beta)}\frac{\partial \hat{\theta}}{\partial y}. \tag{4.23}$$

We can see that $\hat{\phi}$ is triggered only by concentration gradient at this order. We substitute $\hat{\theta} = A(\tau)e^{-\xi^2}$ into the above equation and take the boundary conditions that $\hat{\phi}_1$ is zero as ξ goes to infinity to obtain

$$\hat{\phi}_1 = \frac{\sqrt{\pi}}{2}\alpha^2\tau A(\tau)\frac{D\phi_0}{C_0+\beta}(e^{2\alpha\sqrt{\tau}}e^{\alpha^2\tau}(\mathrm{erf}(\xi+\alpha\sqrt{\tau})-1)+e^{-2\alpha\sqrt{\tau}}e^{\alpha^2\tau}(\mathrm{erf}(\xi-\alpha\sqrt{\tau})+1)).$$

$$(4.24)$$

Therefore, we can estimate $\hat{\phi}$ by

$$\hat{\phi} = \frac{\sqrt{\pi}}{2}\alpha\sqrt{\tau}A(\tau)\frac{D\phi_0}{C_0+\beta}(e^{2\alpha\sqrt{\tau}}e^{\alpha^2\tau}(\mathrm{erf}(\xi+\alpha\sqrt{\tau})-1)+e^{-2\alpha\sqrt{\tau}}e^{\alpha^2\tau}(\mathrm{erf}(\xi-\alpha\sqrt{\tau})+1)).$$

$$(4.25)$$

When $\tau \to \infty$,

$$\mathrm{erf}(\xi+\alpha\sqrt{\tau}) \sim 1 - \frac{e^{(\xi+\alpha\sqrt{\tau})^2}}{(\xi+\alpha\sqrt{\tau})\sqrt{\pi}}\left(1 - \frac{1}{2(\xi+\alpha\sqrt{\tau})^2}\right) \tag{4.26}$$

$$\mathrm{erf}(\xi-\alpha\sqrt{\tau}) \sim 1 - \frac{e^{(\xi-\alpha\sqrt{\tau})^2}}{(\xi-\alpha\sqrt{\tau})\sqrt{\pi}}\left(1 - \frac{1}{2(\xi-\alpha\sqrt{\tau})^2}\right). \tag{4.27}$$

As a result,

$$\hat{\phi} = \frac{D\phi_0}{2(C_0+\beta)}\alpha\sqrt{\tau}\hat{\theta}\left(\frac{-2\xi}{\xi^2-(\alpha\sqrt{\tau})^2} + \frac{2\xi^3+6\xi\alpha^2\tau}{2(\xi^2-\alpha^2\tau)^3}\right)$$

$$\sim \frac{D\phi_0}{C_0+\beta}\hat{\theta}\left(-\frac{1}{\alpha\sqrt{\tau}}+\frac{1}{(\alpha\sqrt{\tau})^3}\right). \tag{4.28}$$

This expression relates the disturbance potential to the disturbance concentration. At large time, $\alpha\sqrt{\tau}\gg 1$, $\phi\sim\frac{\hat{\theta}}{\alpha\sqrt{\tau}}$ and $\hat{\phi}$ is much smaller than $\hat{\theta}$. As a consequence, the potential disturbance $\hat{\phi}$ does not contribute to leading order to the fingering instability and we can omit the third term in equation (4.17).

The final equations are then

$$\frac{\partial^2\hat{\theta}}{\partial\xi^2}+2\xi\frac{\partial\hat{\theta}}{\partial\xi}+2\hat{\theta}=4\tau\frac{\partial\hat{\theta}}{\partial\tau}+4\alpha^2\tau\hat{\theta}+4ae^{-\xi^2}\tau\hat{u} \tag{4.29}$$

$$\frac{\partial^2\hat{u}}{\partial\xi^2}=4\tau\alpha^2\hat{u}+4\sqrt{\tau}Pe\alpha^2 H(\xi)\hat{\theta}. \tag{4.30}$$

Substituting $\hat{\theta}=A(\tau;\alpha)e^{-\xi^2}$ into equation (64) gives:

$$\frac{\partial^2\hat{u}}{\partial\xi^2}=4\tau\alpha^2\hat{u}+4\sqrt{\tau}\alpha^2 PeH(\xi)A(\tau)e^{-\xi^2}. \tag{4.31}$$

The disturbance velocity $\hat{u}$ can be solved explicitly

$$\hat{u}=\frac{1}{4\alpha\sqrt{\tau}}\left[e^{2\alpha\sqrt{\tau}\xi}\left(\int_{-\infty}^{\xi}e^{-2\alpha\sqrt{\tau}\xi}\phi(\xi)d\xi+C_1\right)-e^{-2\alpha\sqrt{\tau}\xi}\left(\int_{-\infty}^{\xi}e^{2\alpha\sqrt{(\tau)}\xi}\phi(\xi)d\xi+C_2\right)\right] \tag{4.32}$$

where $\phi(\xi)=4\sqrt{\tau}\alpha^2 PeH(\xi)A(\tau)e^{-\xi^2}$. The coefficients C_1 and C_2 can be determined by invoking the vanishing boundary conditions of $\hat{u}$ as ξ goes to infinity. The function $H(\xi)$ is a slowly varying function of ξ. Since $H(\xi)=\frac{-I}{2(1+C_0)^{\frac{3}{2}}(C_0+\beta)}$, we can choose $C_0=\frac{c_1+c_2}{2}$ as a good approximation such that H becomes a constant. Finally, we have

$$\hat{u}=\frac{\sqrt{\pi}}{2}\alpha PeHe^{\alpha^2\tau}A(\tau;\alpha)(e^{2\alpha\sqrt{\tau}\xi}(\mathrm{erf}(\xi+\alpha\sqrt{\tau})-1)-e^{-2\alpha\sqrt{\tau}\xi}(\mathrm{erf}(\xi-\alpha\sqrt{\tau})+1)) \tag{4.33}$$

which is of order α.

The solvability condition for a singular equation $L\theta=f$ is that $\langle Lf,\phi\rangle=0$, where ϕ is the corresponding eigenfunction of its adjoint operator L^* and the adjoint operator L^* of L with respect to the L^2 inner product $\langle f,g\rangle=\int_{-\infty}^{+\infty}fgd\xi$ is

$$L^*=\frac{\partial^2}{\partial\xi^2}-2\xi\frac{\partial}{\partial\xi}. \tag{4.34}$$

The null eigenvalue problem $L^*\phi=0$ has a solution $\phi=1$. We then take the inner product of equation (4.29) with 1 to get

$$0=4\tau\langle\frac{\partial\hat{\theta}_0}{\partial\tau},1\rangle+4\alpha^2\tau\langle\hat{\theta}_0,1\rangle+4a\tau\langle\hat{u}e^{-\xi^2},1\rangle. \tag{4.35}$$

Alternatively, this can be written as

$$4\tau \frac{\partial A}{\partial \tau}\sqrt{\pi} + 4\alpha^2 \tau A \sqrt{\pi} + 4a\tau \langle \hat{u}e^{-\xi^2}, 1 \rangle = 0.$$ (4.36)

Expanding the error function as $\tau \to \infty$ gives

$$\hat{u} = \alpha PeH A(\tau; \alpha)\left(-\frac{2e^{-\xi^2}}{\sqrt{\pi}\alpha\sqrt{\tau}} + \frac{e^{-\xi^2}}{\sqrt{\pi}\alpha^3\tau^{\frac{3}{2}}} \right) \cdot \frac{\sqrt{\pi}}{2}.$$ (4.37)

Replacing $\hat{u}$ in equation (4.36), we have

$$\frac{\partial A}{\partial \tau} = \left(-\alpha^2 + \frac{aPeH}{\sqrt{2}\tau^{\frac{1}{2}}} - \frac{aPeH}{2\sqrt{2}\tau^{\frac{3}{2}}\alpha^2} \right) A(\tau; \alpha).$$ (4.38)

A time-dependent growth rate λ hence results:

$$\lambda(\tau; \alpha) = -\alpha^2 + \frac{aPeH}{\sqrt{2}\tau^{\frac{1}{2}}} - \frac{aPeH}{2\sqrt{2}\tau^{\frac{3}{2}}\alpha^2}$$ (4.39)

with a maximum at

$$\alpha_{max} = \left(\frac{aPeH}{2\sqrt{2}} \right)^{\frac{1}{4}} \frac{1}{\tau^{\frac{3}{8}}}.$$ (4.40)

To obtain a more generic form, we define

$$\alpha'_{max} = \frac{\alpha_{max}}{\alpha_0}, \quad \tau' = \tau\tau_0$$ (4.41)

where $\alpha_0 = aPeH\sqrt{\pi}$ and $\tau_0 = (4\sqrt{2}aPeH)^2$, and obtain the universal scaling law

$$\alpha'_{max} = \frac{1.5958}{\tau'^{\frac{3}{8}}}.$$ (4.42)

We have numerically solved equations (4.15), (4.16) and (4.17) for each α and obtain α'_{max} from the simulations. In Figure 3, we successfully collapse the numerical solutions for various values of Pe, β and other parameters into one universal curve that is in good agreement with equation (4.42) at large τ. There is an initial transient of $\tau \sim O(1)$ when other modes, essential and discrete, contribute to the dynamics. However, for $\tau \gg 1$, only the dominant discrete mode survives and its fastest-growing wavenumber coarsens as predicted by equation (4.42).

5 Discussion

The above analysis assumes an unbounded transverse direction and equation (4.40) indicates that, for any given time, there are some disturbances with unrealistically large wavelengths that are unstable. There is actually a physical cutoff due to the finite breadth b of the channel. Using the diffusive length scale L of equation (2.19) to define a dimensionless cutoff wavenumber, we obtain $\alpha_c = \frac{2\pi}{Pe}\frac{w}{b}$, where $\frac{w}{b}$ is the channel aspect ratio.

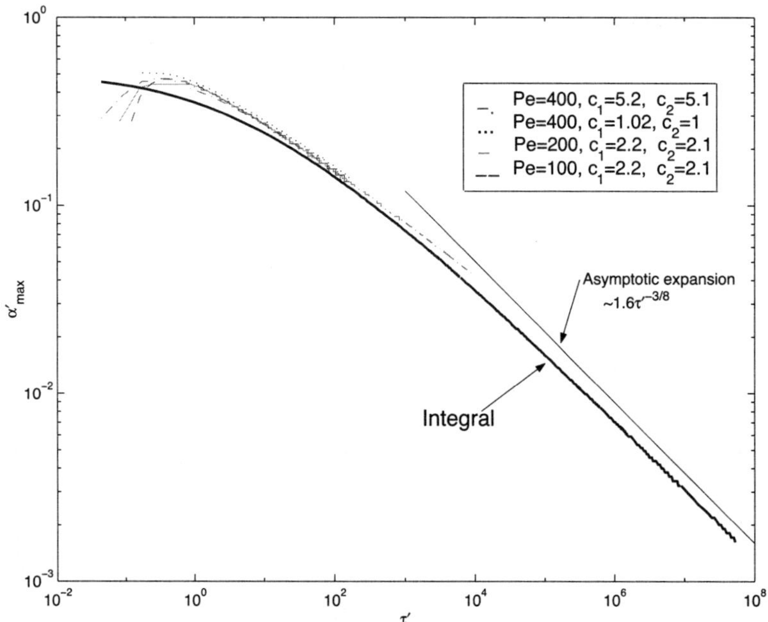

Figure 3. Simulated coarsening rate for indicated values. They collapse into a universal curve in the normalized coordinates of α' and τ' which is in good agreement with the theoretical prediction (4.42). The dotted curves are from a numerical solution of equations (4.15)-(4.17). The solid curve is from a numerical integration of (4.29) and (4.32).

Therefore, the time it takes can also be easily estimated by equaling equation (4.40) and this definition,

$$\tau_c = 3.72 \times 10^{-3} a^{\frac{2}{3}} H^{\frac{2}{3}} Pe^{\frac{2}{3}} \left(\frac{Ub}{D}\right)^{\frac{8}{3}} \tag{5.1}$$

or

$$\tau'_c = 0.1190 \times a^{\frac{8}{3}} Pe^{\frac{8}{3}} H^{\frac{8}{3}} \left(\frac{Ub}{D}\right)^{\frac{8}{3}}. \tag{5.2}$$

One likes to carry out bench-scale experiments in macroscopic devices to understand the dynamics for sub-millimeter microdevices. Such scaling down is especially attractive since the electrokinetic velocity is independent of channel gap width and geometry. Hence, if one uses identical electrolyte concentrations, equations (5.1) and (5.2) indicate that the transient times for two devices with different scales are related by

$$\frac{t_1}{t_2} = \left(\frac{Pe_1}{Pe_2}\right)^{\frac{2}{3}} \left(\frac{U_1}{U_2}\right)^{\frac{2}{3}} \left(\frac{b_1}{b_2}\right)^{\frac{8}{3}} \left(\frac{D_2}{D_1}\right)^{\frac{5}{3}}. \tag{5.3}$$

If the devices are self-similar in geometry and their electrokinetic velocities are identical due to identical applied field, the scaling law is even more explicit:

$$\frac{t_1}{t_2} = \left(\frac{b_1}{b_2}\right)^{\frac{8}{3}}. \tag{5.4}$$

In our experiments shown in Figure 2, this critical time is estimated to be about 20 minutes. For a micro-device, the typical channel width is 500 micron. With other conditions identical to our experiments in Figure 2, equation (5.3) then predicts a fingering transient of $0.4s$ – a negligible value. In contrast, any sample in a micro-device has a front that has diffused for more than one minute – roughly the residence time for most devices. As a result, the fingering transient has long expired and one does not expect miscible fingering to be an issue in micro-devices.

Acknowledgement

This work was supported by an XYZ-on-a-Chip grant from NSF.

Bibliography

R. Aris. On the dispersion of a solute in a fluid flowing through a tube. *Proc. Roy. Soc. London Ser. A*, 235:67–77, 1956.

H.-C. Chang and E.A. Demekhin. *Complex Wave Dynamics on Thin Films*. Elsevier, New York, 2002.

C.T. Culbertson, S.C. Jacobson, and J.M. Ramsey. Dispersion sources for compact geometries in microchips. *Anal. Chem.*, 70:3781–3789, 1998.

S.V. Ermakov, S.C. Jacobson, and J.M. Ramsey. Computer simulations of electrokinetic transport in microfabricated channel structures. *Anal. Chem.*, 70:4494–4504, 1998.

M. Freemantle. Downsizing chemistry. *Chemical & Engineering News*, 77:27–36, 1999.

A.E. Herr, J.I. Molho, J.G. Santiago, M.G. Mulgal, and T.W. Kenny. Electroosmotic capillary flow with nonuniform zeta potential. *Anal. Chem.*, 72:1053–1057, 2000.

H.H. Hu and D.D. Joseph. Miscible displacement in a Hele-Shaw cell. *ZAMP*, 43:626–644, 1992.

R.F. Probstein. *Physiochemical Hydrodynamics*. John Wiley & Sons, Inc., 1994.

P.V. Takhistov, A. Indeikina, and H.-C. Chang. Electrokinetic displacement of air bubbles in micro-channels. *Phys. Fluids*, 14:1–14, 2002.

C.T. Tan and G.M. Homsy. Stability of miscible displacements in porous media: Rectilinear flow. *Phys. Fluids*, 29:3549–3556, 1986.

G.I. Taylor. The dispersion of matter in turbulent flow through a pipe. *Proc. Roy. Soc. London Ser. A*, 223:446–468, 1954.

Y.C. Yortsos and M. Zeybek. Dispersion driven instability in miscible displacement in porous media. *Phys. Fluids*, 31:3511–3518, 1988.